DIFFERENTIAL EQUATIONS: THEORY AND APPLICATIONS

Edited by
Maria Catherine C. Borres

ArclerPress

Differential Equations: Theory and Applications

Edited by Maria Catherine C. Borres

Edition 2018

ArclerPress

Published by
Arcler Press LLC
708 3rd Avenue, 6th Floor
New York
NY 10017
United States of America
www.arclerpress.com

ISBN: 978-1-68094-589-8 (Hardback)

Printed in United States of America

Library of Congress Control Number: 2017933178

About the Editor

Maria Catherine C. Borres

Catherine is currently taking up Master of Arts in Education Major in Mathematics in Philippine Normal University – Manila. She is currently working as a Content Developer for Mathematics at the Affordable Private Education Center (APEC Schools).

Table of Contents

2 ORDINARY DIFFERENTIAL EQUATIONS

3 PARTIAL DIFFERENTIAL EQUATIONS

4 DIFFERENTIAL ALGEBRAIC EQUATIONS

5 LINEAR DIFFERENTIAL EQUATIONS

6 DELAY DIFFERENTIAL EQUATIONS

7 AUTONOMOUS DIFFERENTIAL EQUATIONS

8 STOCHASTIC DIFFERENTIAL EQUATIONS

List of Abbreviations

DDE	Delay Differential Equation
DAE	Differential Algebraic Equation
ODE	Ordinary Differential Equation
DGFEM	Discontinuous Galerkin Finite Element Method
EFGM	Element-Free Galerkin Method
FEM	Finite Element Method
FDE	Functional Differential Equation
GFEM	Generalized Finite Element Method
SPDEs	Stochastic Partial Differential Equations
SDE	Stochastic Differential Equation
SPPS	Spectral Parameter Power Series
TIV	Time-Invariant

Preface

A differential equation is a mathematical equation that relates some function with its derivatives. In applications, the functions usually represent physical quantities, the derivatives represent their rates of change, and the equation defines a relationship between the two. Because such relations are extremely common, differential equations play a prominent role in many disciplines including engineering, physics, economics, and biology. In pure mathematics, differential equations are studied from several different perspectives, mostly concerned with their solutions—the set of functions that satisfy the equation. Only the simplest differential equations are solvable by explicit formulas; however, some properties of solutions of a given differential equation may be determined without finding their exact form. Differential equations can be divided into several types. Apart from describing the properties of the equation itself, these classes of differential equations can help inform the choice of approach to a solution. Commonly used distinctions include whether the equation is: Ordinary/Partial, Linear/Non-linear, and Homogeneous/Inhomogeneous. One of the most basic examples of differential equations is the Malthusian Law of population growth $dp/dt = rp$ shows how the population (p) changes with respect to time. The constant r will change depending on the species. Malthus used this law to predict how a species would grow over time. More complicated differential equations can be used to model the relationship between predators and prey. For example, as predators increase then prey decrease as more get eaten. But then the predators will have less to eat and start to die out, which allows more prey to survive. The interactions between the two populations are connected by differential equations.

Content Coverage

Chapters One and *Two* present an introduction of differential equations and ordinary differential equations, correspondingly. Ordinary differential equations (ODEs) arise in many contexts of mathematics and science.

Chapters Three and *Four* present a coverage of partial differential equation (PDE) and differential-algebraic equation, respectively. PDEs can be used to describe a wide variety of phenomena such as sound, heat, electrostatics, electrodynamics, fluid dynamics, elasticity, or quantum mechanics.

Chapter Five presents emphasis on linear differential equations which are differential equations having solutions which can be added together in particular linear combinations to form further solutions.

Chapter Six focuses on delay differential equations to be useful in various fields such as age-structured population growth, control theory, and any model involving responses with non-zero delays.

Chapters Seven and *Eight* focus on autonomous differential equations and post-translational modification of proteins, separately. Autonomous systems are closely related to dynamical systems. Any autonomous system can be transformed into a dynamical system and, using very weak assumptions, a dynamical system can be transformed into an autonomous system.

CHAPTER 1

Differential Equation: An Overview

Chapter Contents

- **Introduction**
- **History**
- **Types**
- **Existence of Solutions**
- **Connection to Difference Equations**
- **Applications**
- **Complex Differential Equation**
- **Exact Differential**
- **Inexact Differential Equation**
- **Closed and Exact Differential Forms**

By studying the exponential functions, and picking a convenient base, we have inadvertently stumbled on a relationship satisfied by the function and its derivative. That relationship is called a differential equation. The topic of differential equations is an extremely important one in mathematics and science, as well as many other branches of studies (economics, commerce) in which changes occur and in which predictions are desirable. In most such circumstances, the systems studied come with some kind of "Natural Laws", or observations that, when translated into the language of mathematics, become differential equations. It is then the job of the mathematician to try to figure out what are the predictions, i.e. to find the functions that satisfy those equations.

INTRODUCTION

A differential equation is a mathematical equation that relates some function with its derivatives. In applications, the functions usually represent physical quantities, the derivatives represent their rates of change, and the equation defines a relationship between the two. Because such relations are extremely common, differential equations play a prominent role in many disciplines including engineering, physics, economics, and biology.

In pure mathematics, differential equations are studied from several different perspectives, mostly concerned with their solutions—the set of functions that satisfy the equation. Only the simplest differential equations are solvable by explicit formulas; however, some properties of solutions of a given differential equation may be determined without finding their exact form.

If a self-contained formula for the solution is not available, the solution may be numerically approximated using computers. The theory of dynamical systems puts emphasis on qualitative analysis of systems described by differential equations, while many numerical methods have been developed to determine solutions with a given degree of accuracy.

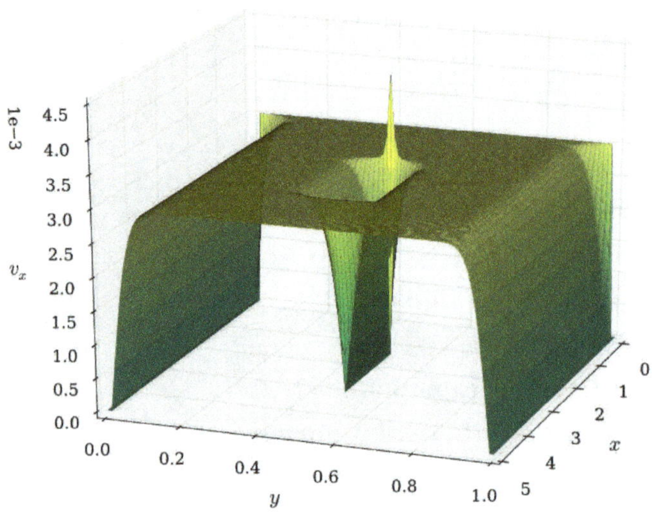

Navier–Stokes differential equations used to simulate airflow around an obstruction.

HISTORY

Differential equations first came into existence with the invention of calculus by Newton and Leibniz. In Chapter 2 of his 1671 work "Methodus fluxionum et Serierum Infinitarum", Isaac Newton listed three kinds of differential equations:

$$\frac{dy}{dx} = f(x)$$

$$\frac{dy}{dx} = f(x, y)$$

$$x_1 \frac{\partial y}{\partial x_1} + x_2 \frac{\partial y}{\partial x_2} = y$$

He solves these examples and others using infinite series and discusses the non-uniqueness of solutions.

Jacob Bernoulli proposed the Bernoulli differential equation in 1695. This is an ordinary differential equation of the form

$$y' + P(x)y = Q(x)y^n$$

for which the following year Leibniz obtained solutions by simplifying it.

Historically, the problem of a vibrating string such as that of a musical instrument was studied by Jean le Rond d'Alembert, Leonhard Euler, Daniel Bernoulli, and Joseph-Louis Lagrange. In 1746, d'Alembert discovered the one-dimensional wave equation, and within ten years Euler discovered the three-dimensional wave equation.

The Euler–Lagrange equation was developed in the 1750s by Euler and Lagrange in connection with their studies of the tautochrone problem. This is the problem of determining a curve on which a weighted particle will fall to a fixed point in a fixed amount of time, independent of the starting point.

Lagrange solved this problem in 1755 and sent the solution to Euler. Both further developed Lagrange's method and applied it to mechanics, which led to the formulation of Lagrangian mechanics.

Fourier published his work on heat flow in Théorie analytique de la chaleur (The Analytic Theory of Heat), in which he based his reasoning on Newton's law of cooling, namely, that the flow of heat between two adjacent molecules is proportional to the extremely small difference of their temperatures. Contained in this book was Fourier's proposal of his heat equation for conductive diffusion of heat. This partial differential equation is now taught to every student of mathematical physics.

Example

For example, in classical mechanics, the motion of a body is described by its position and velocity as the time value varies. Newton's laws allow (given the position, velocity, acceleration and various forces acting on the body) one to express these variables dynamically as a differential equation for the unknown position of the body as a function of time.

In some cases, this differential equation (called an equation of motion) may be solved explicitly.

An example of modelling a real world problem using differential equations is the determination of the velocity of a ball falling through the air, considering only gravity and air resistance. The ball's acceleration towards the ground is the acceleration due to gravity minus the acceleration due to air resistance.

Gravity is considered constant, and air resistance may be modeled as proportional to the ball's velocity. This means that the ball's acceleration, which is a derivative of its velocity, depends on the velocity (and the velocity depends on time). Finding the velocity as a function of time involves solving a differential equation and verifying its validity.

TYPES

Differential equations can be divided into several types. Apart from describing the properties of the equation itself, these classes of differential equations can help inform the choice of approach to a solution. Commonly used distinctions include whether the equation is: Ordinary/Partial, Linear/Non-linear, and Homogeneous/Inhomogeneous. This list is far from exhaustive; there are many other properties and subclasses of differential equations which can be very useful in specific contexts.

Ordinary Differential Equations

An ordinary differential equation (ODE) is an equation containing a function of one independent variable and its derivatives. The term "ordinary" is used in contrast with the term partial differential equation which may be with respect to more than one independent variable.

Linear differential equations, which have solutions that can be added and multiplied by coefficients, are well-defined and understood, and exact closed-form solutions are obtained. By contrast, ODEs that lack additive solutions are nonlinear, and solving them is far more intricate, as one can rarely represent them by elementary functions in closed form: Instead, exact and analytic solutions of ODEs are in series or integral form. Graphical and numerical methods, applied by hand or by computer, may approximate solutions of ODEs and perhaps yield useful information, often sufficing in the absence of exact, analytic solutions.

Partial Differential Equations

A partial differential equation (PDE) is a differential equation that contains unknown multivariable functions and their partial derivatives. (This is in contrast to ordinary differential equations, which deal with functions of a single variable and their derivatives.) PDEs are used to formulate problems involving functions of several variables, and are either solved in closed form, or used to create a relevant computer model.

PDEs can be used to describe a wide variety of phenomena such as sound, heat, electrostatics, electrodynamics, fluid flow, elasticity, or quantum mechanics. These seemingly distinct physical phenomena can be formalised similarly in terms of PDEs. Just as ordinary differential equations often model one-dimensional dynamical systems, partial differential equations often model multidimensional systems. PDEs find their generalisation in stochastic partial differential equations.

Linear Differential Equations

A differential equation is linear if the unknown function and its derivatives have degree 1 (products of the unknown function and its derivatives are not allowed) and nonlinear otherwise. The characteristic property of linear equations is that their solutions form an affine subspace of an appropriate function space, which results in much more developed theory of linear differential equations.

Homogeneous linear differential equations are a subclass of linear differential equations for which the space of solutions is a linear subspace i.e. the sum of any set of solutions or multiples of solutions is also a solution. The coefficients of the unknown function and its derivatives in a linear differential equation are allowed to be (known) functions of the independent variable or variables; if these coefficients are constants then one speaks of a constant coefficient linear differential equation.

Non-Linear Differential Equations

Non-linear differential equations are formed by the products of the unknown function and its derivatives are allowed and its degree is > 1.There are very few methods of solving nonlinear differential equations exactly; those that are known typically depend on the equation having particular symmetries. Nonlinear differential equations can exhibit very complicated behavior over extended time intervals, characteristic of chaos. Even the fundamental questions of existence, uniqueness, and extendability of solutions for nonlinear differential equations, and well-posedness of initial and boundary value problems for nonlinear PDEs are hard problems and their resolution in special cases is considered to be a significant advance in the mathematical theory (cf. Navier–Stokes existence and smoothness). However, if the differential equation is a correctly formulated representation of a meaningful physical process, then one expects it to have a solution.

Linear differential equations frequently appear as approximations to nonlinear equations. These approximations are only valid under restricted conditions. For example, the harmonic oscillator equation is an approximation to the nonlinear pendulum equation that is valid for small amplitude oscillations (see below).

Equation Order

Differential equations are described by their order, determined by the term with the highest derivatives. An equation containing only first derivatives is a first-order differential equation, an equation containing the second derivative is a second-order differential equation, and so on.

Examples

In the first group of examples, let u be an unknown function of x, and c and ω are known constants. Note both ordinary and partial differential equations are broadly classified as linear and nonlinear.

- Inhomogeneous first-order linear constant coefficient ordinary differential equation:

$$\frac{du}{dx} = cu + x^2.$$

- Homogeneous second-order linear ordinary differential equation:

$$\frac{d^2u}{dx^2} - x\frac{du}{dx} + u = 0.$$

- Homogeneous second-order linear constant coefficient ordinary differential equation describing the harmonic oscillator:

$$\frac{d^2u}{dx^2} + \omega^2 u = 0.$$

- Inhomogeneous first-order nonlinear ordinary differential equation:

$$\frac{du}{dx} = u^2 + 4.$$

- Second-order nonlinear (due to sine function) ordinary differential equation describing the motion of a pendulum of length L:

$$L\frac{d^2u}{dx^2} + g\sin u = 0.$$

In the next group of examples, the unknown function u depends on two variables x and t or x and y.

- Homogeneous first-order linear partial differential equation:

$$\frac{\partial u}{\partial t} + t\frac{\partial u}{\partial x} = 0.$$

- Homogeneous second-order linear constant coefficient partial differential equation of elliptic type, the Laplace equation:

$$\frac{\partial^2 u}{\partial x^2} + \frac{\partial^2 u}{\partial y^2} = 0.$$

$$\frac{\partial u}{\partial t} = 6u\frac{\partial u}{\partial x} - \frac{\partial^3 u}{\partial x^3}.$$

EXISTENCE OF SOLUTIONS

Solving differential equations is not like solving algebraic equations. Not only are their solutions oftentimes unclear, but whether solutions are unique or exist at all are also notable subjects of interest.

For first order initial value problems, the Peano existence theorem gives one set of circumstances in which a solution exists. Given any point (a, b) in the xy-plane, define some rectangular region Z, such that $Z = [l, m] \times [n, p]$ and (a, b) is in the interior of Z. If we are given a differential equation $\frac{dy}{dx} = g(x, y)$ and the condition that $y = b$ when $x = a$, then there is locally a solution to this problem if $g(x, y)$ and $\frac{\partial g}{\partial x}$ are both continuous on Z. This solution exists on some interval with its center at a. The solution may not be unique. (

However, this only helps us with first order initial value problems. Suppose we had a linear initial value problem of the nth order:

$$f_n(x)\frac{d^n y}{dx^n} + \cdots + f_1(x)\frac{dy}{dx} + f_0(x)y = g(x)$$

such that

$$y(x_0) = y_0, y'(x_0) = y'_0, y''(x_0) = y''_0, \cdots$$

For any nonzero $f_n(x)$, if $\{f_0, f_1, \cdots\}$ and g are continuous on some interval containing x_0, y is unique and exists.

Related Concepts

- A delay differential equation (DDE) is an equation for a function of a single variable, usually called time, in which the derivative of the function at a certain time is given in terms of the values of the function at earlier times.
- A stochastic differential equation (SDE) is an equation in which the unknown quantity is a stochastic process and the equation involves some known stochastic processes, for example, the Wiener process in the case of diffusion equations.
- A differential algebraic equation (DAE) is a differential equation comprising differential and algebraic terms, given in implicit form.

CONNECTION TO DIFFERENCE EQUATIONS

The theory of differential equations is closely related to the theory of difference equations, in which the coordinates assume only discrete values, and the relationship involves values of the unknown function or functions and values at nearby coordinates. Many methods to compute numerical solutions of differential equations or study the properties of differential equations involve approximation of the solution of a differential equation by the solution of a corresponding difference equation.

APPLICATIONS

The study of differential equations is a wide field in pure and applied mathematics, physics, and engineering. All of these disciplines are concerned with the properties of differential equations of various types. Pure mathematics focuses on the existence and uniqueness of solutions, while applied mathematics emphasizes the rigorous justification of the methods for approximating solutions. Differential equations play an important role in modelling virtually every physical, technical, or biological process, from celestial motion, to bridge design, to interactions between neurons. Differential equations such as those used to solve real-life problems may not necessarily be directly solvable, i.e. do not have closed form solutions. Instead, solutions can be approximated using numerical methods.

Many fundamental laws of physics and chemistry can be formulated as differential equations. In biology and economics, differential equations are used to model the behavior of complex systems. The mathematical theory of differential equations first developed together with the sciences where the equations had originated and where the results found application. However, diverse problems, sometimes originating in quite distinct scientific fields, may give rise to identical differential equations. Whenever this happens, mathematical theory behind the equations can be viewed as a unifying principle behind diverse phenomena. As an example, consider propagation of light and sound in the atmosphere, and of waves on the surface of a pond. All of them may be described by the same second-order partial differential equation, the wave equation, which allows us to think of light and sound as forms of waves, much like familiar waves in the water. Conduction of heat, the theory of which was developed by Joseph Fourier, is governed by another second-order partial differential equation, the heat equation. It turns out that many diffusion processes, while seemingly different, are described by the same equation; the Black–Scholes equation in finance is, for instance, related to the heat equation.

Physics

- Euler–Lagrange equation in classical mechanics
- Hamilton's equations in classical mechanics
- Radioactive decay in nuclear physics
- Newton's law of cooling in thermodynamics
- The wave equation
- The heat equation in thermodynamics
- Laplace's equation, which defines harmonic functions
- Poisson's equation
- The geodesic equation
- The Navier–Stokes equations in fluid dynamics
- The Diffusion equation in stochastic processes
- The Convection–diffusion equation in fluid dynamics
- The Cauchy–Riemann equations in complex analysis
- The Poisson–Boltzmann equation in molecular dynamics
- The shallow water equations
- Universal differential equation
- The Lorenz equations whose solutions exhibit chaotic flow.

Classical Mechanics

So long as the force acting on a particle is known, Newton's second law is sufficient to describe the motion of a particle. Once independent relations for each force acting on a particle are available, they can be substituted into Newton's second law to obtain an ordinary differential equation, which is called the equation of motion.

Electrodynamics

Maxwell's equations are a set of partial differential equations that, together with the Lorentz force law, form the foundation of classical electrodynamics, classical optics, and electric circuits. These fields in turn underlie modern electrical and communications technologies. Maxwell's equations describe how electric and magnetic fields are generated and altered by each other and by charges and currents. They are named after the Scottish physicist and mathematician James Clerk Maxwell, who published an early form of those equations between 1861 and 1862.

General Relativity

The Einstein field equations (EFE; also known as "Einstein's equations") are a set of ten partial differential equations in Albert Einstein's general theory of relativity which describe the fundamental interaction of gravitation as a result of spacetime being curved by matter and energy. First published by Einstein in 1915 as a tensor equation, the EFE equate local spacetime curvature (expressed by the Einstein tensor) with the local energy and momentum within that spacetime (expressed by the stress–energy tensor).

Quantum Mechanics

In quantum mechanics, the analogue of Newton's law is Schrödinger's equation (a partial differential equation) for a quantum system (usually atoms, molecules, and subatomic particles whether free, bound, or localized). It is not a simple algebraic equation, but in general a linear partial differential equation, describing the time-evolution of the system's wave function (also called a "state function").

Biology

- Verhulst equation – biological population growth
- von Bertalanffy model – biological individual growth
- Replicator dynamics – found in theoretical biology
- Hodgkin–Huxley model – neural action potentials

Predator-Prey Equations

The Lotka–Volterra equations, also known as the predator–prey equations, are a pair of first-order, non-linear, differential equations frequently used to describe the dynamics of biological systems in which two species interact, one as a predator and the other as prey.

Chemistry

The rate law or rate equation for a chemical reaction is a differential equation that links the reaction rate with concentrations or pressures of reactants and constant parameters (normally rate coefficients and partial reaction orders). To determine the rate equation for a particular system one combines the reaction rate with a mass balance for the system.

Economics

- The key equation of the Solow–Swan model is $\frac{\partial k(t)}{\partial t} = s[k(t)]^{\alpha} - \delta k(t)$
- The Black–Scholes PDE
- Malthusian growth model
- The Vidale–Wolfe advertising model

COMPLEX DIFFERENTIAL EQUATION

A complex differential equation is a differential equation whose solutions are functions of a complex variable.

Constructing integrals involves choice of what path to take, which means singularities and branch points of the equation need to be studied. Analytic continuation is used to generate new solutions and this means topological considerations such as monodromy, coverings and connectedness are to be taken into account.

Existence and uniqueness theorems involve the use of majorants and minorants.

Study of rational second order ODEs in the complex plane led to the discovery of new transcendental special functions which are now known as Painlevé transcendents.

Nevanlinna theory can be used to study complex differential equations. This leads to extensions of Malmquist's theorem.

Generalizations

Generalizations include partial differential equations in several complex variables, or differential equations on complex manifolds. Also there are at least a couple of ways of studying complex difference equations: either study holomorphic functions which satisfy functional relations given by the difference equation or study discrete analogs of holomorphicity such as monodiffric functions. Also integral equations can be studied in the complex domain.

EXACT DIFFERENTIAL

In multivariate calculus, a differential is said to be exact or perfect, as contrasted with an inexact differential, if it is of the form dQ, for some differentiable function Q.

Overview

Definition

We work in three dimensions, with similar definitions holding in any other number of dimensions. In three dimensions, a form of the type

$$A(x, y, z)\,dx + B(x, y, z)\,dy + C(x, y, z)\,dz$$

is called a differential form. This form is called exact on a domain $D \subset \mathbb{R}^3$ in space if there exists some scalar function $Q = Q(x, y, z)$ defined on D such that

$$dQ \equiv \left(\frac{\partial Q}{\partial x}\right)_{y,z} dx + \left(\frac{\partial Q}{\partial y}\right)_{x,z} dy + \left(\frac{\partial Q}{\partial z}\right)_{x,y} dz, \qquad dQ = A\,dx + B\,dy + C\,dz$$

throughout D. This is equivalent to saying that the vector field (A, B, C) is a conservative vector field, with corresponding potential Q.

Note: The subscripts outside the parenthesis indicate which variables are being held constant during differentiation. Due to the definition of the partial derivative, these subscripts are not required, but they are included as a reminder.

One Dimension

In one dimension, a differential form

$$A(x)\,dx$$

is exact as long as a has an antiderivative (but not necessarily one in terms of elementary functions). If A has an antiderivative, let Q be an antiderivative of A and this Q satisfies the condition for exactness. If A does not have an antiderivative, we cannot write $dQ = A(x)\,dx$ and so the differential form is inexact.

Two and Three Dimensions

By symmetry of second derivatives, for any "nice" (non-pathological) function Q we have

$$\frac{\partial^2 Q}{\partial x\,\partial y} = \frac{\partial^2 Q}{\partial y\,\partial x}$$

Hence, it follows that in a simply-connected region R of the xy-plane, a differential

$$A(x, y)\,dx + B(x, y)\,dy$$

is an exact differential if and only if the following holds:

$$\left(\frac{\partial A}{\partial y}\right)_x = \left(\frac{\partial B}{\partial x}\right)_y$$

For three dimensions, a differential

$$dQ = A(x, y, z)\, dx + B(x, y, z)\, dy + C(x, y, z)\, dz$$

is an exact differential in a simply-connected region R of the xyz-coordinate system if between the functions A, B and C there exist the relations:

$$\left(\frac{\partial A}{\partial y}\right)_{x,z} = \left(\frac{\partial B}{\partial x}\right)_{y,z} ; \left(\frac{\partial A}{\partial z}\right)_{x,y} = \left(\frac{\partial C}{\partial x}\right)_{y,z} ; \left(\frac{\partial B}{\partial z}\right)_{x,y} = \left(\frac{\partial C}{\partial y}\right)_{x,z}$$

These conditions are equivalent to the following one: If G is the graph of this vector valued function then for all tangent vectors X,Y of the surface G then s(X, Y) = 0 with s the symplectic form.

These conditions, which are easy to generalize, arise from the independence of the order of differentiations in the calculation of the second derivatives. So, in order for a differential dQ, that is a function of four variables to be an exact differential, there are six conditions to satisfy.

In summary, when a differential dQ is exact:

- the function Q exists;
- $\int_i^f dQ = Q(f) - Q(i)$, independent of the path followed.

In thermodynamics, when dQ is exact, the function Q is a state function of the system. The thermodynamic functions U, S, H, A and G are state functions. Generally, neither work nor heat is a state function. An exact differential is sometimes also called a 'total differential', or a 'full differential', or, in the study of differential geometry, it is termed an exact form.

Partial Differential Relations

If three variables, x, y and z are bound by the condition $F(x, y, z) = \text{constant}$ for some differentiable function $F(x, y, z)$, then the following total differentials exist

$$dx = \left(\frac{\partial x}{\partial y}\right)_z dy + \left(\frac{\partial x}{\partial z}\right)_y dz$$

$$dz = \left(\frac{\partial z}{\partial x}\right)_y dx + \left(\frac{\partial z}{\partial y}\right)_x dy.$$

Substituting the first equation into the second and rearranging, we obtain

$$dz = \left(\frac{\partial z}{\partial x}\right)_y \left[\left(\frac{\partial x}{\partial y}\right)_z dy + \left(\frac{\partial x}{\partial z}\right)_y dz\right] + \left(\frac{\partial z}{\partial y}\right)_x dy,$$

$$dz = \left[\left(\frac{\partial z}{\partial x}\right)_y \left(\frac{\partial x}{\partial y}\right)_z + \left(\frac{\partial z}{\partial y}\right)_x\right] dy + \left(\frac{\partial z}{\partial x}\right)_y \left(\frac{\partial x}{\partial z}\right)_y dz,$$

$$\left[1 - \left(\frac{\partial z}{\partial x}\right)_y \left(\frac{\partial x}{\partial z}\right)_y\right] dz = \left[\left(\frac{\partial z}{\partial x}\right)_y \left(\frac{\partial x}{\partial y}\right)_z + \left(\frac{\partial z}{\partial y}\right)_x\right] dy.$$

Since y and z are independent variables dy and dz may be chosen without restriction. For this last equation to hold in general, the bracketed terms must be equal to zero.

Reciprocity Relation

Setting the first term in brackets equal to zero yields

$$\left(\frac{\partial z}{\partial x}\right)_y \left(\frac{\partial x}{\partial z}\right)_y = 1.$$

A slight rearrangement gives a reciprocity relation,

$$\left(\frac{\partial z}{\partial x}\right)_y = \frac{1}{\left(\frac{\partial x}{\partial z}\right)_y}.$$

There are two more permutations of the foregoing derivation that give a total of three reciprocity relations between x, y and z. Reciprocity relations show that the inverse of a partial derivative is equal to its reciprocal.

Cyclic Relation

The cyclic relation is also known as the cyclic rule or the Triple product rule. Setting the second term in brackets equal to zero yields

$$\left(\frac{\partial z}{\partial x}\right)_y \left(\frac{\partial x}{\partial y}\right)_z = -\left(\frac{\partial z}{\partial y}\right)_x.$$

Using a reciprocity relation for $\frac{\partial z}{\partial y}$ on this equation and reordering gives a cyclic relation (the triple product rule),

$$\left(\frac{\partial x}{\partial y}\right)_z \left(\frac{\partial y}{\partial z}\right)_x \left(\frac{\partial z}{\partial x}\right)_y = -1.$$

If, instead, a reciprocity relation for $\frac{\partial x}{\partial y}$ is used with subsequent rearrangement, a standard form for implicit differentiation is obtained:

$$\left(\frac{\partial y}{\partial x}\right)_z = -\frac{\left(\frac{\partial z}{\partial x}\right)_y}{\left(\frac{\partial z}{\partial y}\right)_x}.$$

Some Useful Equations Derived From Exact Differentials in Two Dimensions

Suppose we have five state functions $z, x, y, u,$ and v. Suppose that the state space is two dimensional and any of the five quantities are exact differentials. Then by the chain rule

$$(1)\quad dz = \left(\frac{\partial z}{\partial x}\right)_y dx + \left(\frac{\partial z}{\partial y}\right)_x dy = \left(\frac{\partial z}{\partial u}\right)_v du + \left(\frac{\partial z}{\partial v}\right)_u dv$$

but also by the chain rule:

$$(2)\quad dx = \left(\frac{\partial x}{\partial u}\right)_v du + \left(\frac{\partial x}{\partial v}\right)_u dv$$

and

$$(3)\quad dy = \left(\frac{\partial y}{\partial u}\right)_v du + \left(\frac{\partial y}{\partial v}\right)_u dv$$

so that:

$$(4)\quad dz = \left[\left(\frac{\partial z}{\partial x}\right)_y \left(\frac{\partial x}{\partial u}\right)_v + \left(\frac{\partial z}{\partial y}\right)_x \left(\frac{\partial y}{\partial u}\right)_v\right] du + \left[\left(\frac{\partial z}{\partial x}\right)_y \left(\frac{\partial x}{\partial v}\right)_u + \left(\frac{\partial z}{\partial y}\right)_x \left(\frac{\partial y}{\partial v}\right)_u\right] dv$$

which implies that:

$$(5)\quad \left(\frac{\partial z}{\partial u}\right)_v = \left(\frac{\partial z}{\partial x}\right)_y \left(\frac{\partial x}{\partial u}\right)_v + \left(\frac{\partial z}{\partial y}\right)_x \left(\frac{\partial y}{\partial u}\right)_v$$

Letting $v = y$ gives:

$$(6)\quad \left(\frac{\partial z}{\partial u}\right)_y = \left(\frac{\partial z}{\partial x}\right)_y \left(\frac{\partial x}{\partial u}\right)_y$$

Letting $u = y$ gives:

$$(7)\quad \left(\frac{\partial z}{\partial y}\right)_v = \left(\frac{\partial z}{\partial y}\right)_x + \left(\frac{\partial z}{\partial x}\right)_y \left(\frac{\partial x}{\partial y}\right)_v$$

Letting $u = y$, $v = z$ gives:

$$(8)\quad \left(\frac{\partial z}{\partial y}\right)_x = -\left(\frac{\partial z}{\partial x}\right)_y \left(\frac{\partial x}{\partial y}\right)_z$$

using $(\partial a/\partial b)_c = 1/(\partial b/\partial a)_c$ gives the triple product rule:

$$\text{(9)} \quad \left(\frac{\partial z}{\partial x}\right)_y \left(\frac{\partial x}{\partial y}\right)_z \left(\frac{\partial y}{\partial z}\right)_x = -1$$

Exact Differential Equation

In mathematics, an exact differential equation or total differential equation is a certain kind of ordinary differential equation which is widely used in physics and engineering.

Definition

Given a simply connected and open subset D of $\mathbf{R}^2$ and two functions I and J which are continuous on D then an implicit first-order ordinary differential equation of the form

$$I(x, y)\,\mathrm{d}x + J(x, y)\,\mathrm{d}y = 0,$$

is called an exact differential equation if there exists a continuously differentiable function F, called the potential function, so that

$$\frac{\partial F}{\partial x} = I$$

and

$$\frac{\partial F}{\partial y} = J.$$

The nomenclature of "exact differential equation" refers to the exact derivative of a function. For a function $F(x_0, x_1, \ldots, x_{n-1}, x_n)$, the exact or total derivative with respect to 0 is given by

$$\frac{\mathrm{d}F}{\mathrm{d}x_0} = \frac{\partial F}{\partial x_0} + \sum_{i=1}^{n} \frac{\partial F}{\partial x_i}\frac{\mathrm{d}x_i}{\mathrm{d}x_0}.$$

Example

The function $F : \mathbb{R}^2 \to \mathbb{R}$ given by

$$F(x, y) = \frac{1}{2}(x^2 + y^2)$$

is a potential function for the differential equation

$$x\,\mathrm{d}x + y\,\mathrm{d}y = 0.$$

Existence of Potential Functions

In physical applications the functions I and J are usually not only continuous but even continuously differentiable. Schwarz's Theorem then provides us with a necessary criterion for the existence of a potential function. For differential equations defined on simply connected sets the criterion is even sufficient and we get the following theorem:

Given a differential equation of the form (for example, when F has zero slope in the x and y direction at F(x,y)):

$$I(x, y)\, \mathrm{d}x + J(x, y)\, \mathrm{d}y = 0,$$

with I and J continuously differentiable on a simply connected and open subset D of $\mathbf{R}^2$ then a potential function F exists if and only if

$$\frac{\partial I}{\partial y}(x, y) = \frac{\partial J}{\partial x}(x, y).$$

Solutions to Exact Differential Equations

Given an exact differential equation defined on some simply connected and open subset D of $\mathbf{R}^2$ with potential function F then a differentiable function f with (x, f(x)) in D is a solution if and only if there exists real number c so that

$$F(x, f(x)) = c.$$

For an initial value problem

$$y(x_0) = y_0$$

we can locally find a potential function by

$$F(x, y) = \int_{x_0}^{x} I(t, y_0)\mathrm{d}t + \int_{y_0}^{y} J(x, t)\mathrm{d}t = \int_{x_0}^{x} I(t, y_0)\mathrm{d}t + \int_{y_0}^{y} \left[J(x_0, t) + \int_{x_0}^{x} \frac{\partial I}{\partial t}(u, t)\, \mathrm{d}u \right] \mathrm{d}t.$$

Solving

$$F(x, y) = c$$

for y, where c is a real number, we can then construct all solutions.

INEXACT DIFFERENTIAL EQUATION

An inexact differential equation is a differential equation of the form

$$M(x, y)\, dx + N(x, y)\, dy = 0, \text{ where } \frac{\partial M}{\partial y} \neq \frac{\partial N}{\partial x}.$$

The solution to such equations came with the invention of the integrating factor by Leonhard Euler in 1739.

Solution Method

In order to solve the equation, we need to transform it into an exact differential equation. in order to do that, we need to find an integrating factor μ to multiply the equation by. We'll start with the equation itself.
$M\,dx + N\,dy = 0$, so we get $\mu M\,dx + \mu N\,dy = 0$. We will require μ to satisfy $\frac{\partial \mu M}{\partial y} = \frac{\partial \mu N}{\partial x}$. We get
$\frac{\partial \mu}{\partial y}M + \frac{\partial M}{\partial y}\mu = \frac{\partial \mu}{\partial x}N + \frac{\partial N}{\partial x}\mu$. After simplifying we get $M\mu_y - N\mu_x + (M_y - N_x)\mu = 0$. Since this is a partial differential equation, it is mostly extremely hard to solve, however in most cases we will get either $\mu(x,y) = \mu(x)$ or $\mu(x,y) = \mu(y)$, in which case we only need to find μ with a first-order linear differential equation or a separable differential equation, and as such either $\mu(y) = e^{-\int \frac{M_y - N_x}{M}\,dy}$ or $\mu(x) = e^{\int \frac{M_y - N_x}{N}\,dx}$.

CLOSED AND EXACT DIFFERENTIAL FORMS

In mathematics, especially vector calculus and differential topology, a closed form is a differential form α whose exterior derivative is zero (dα = 0), and an exact form is a differential form that is the exterior derivative of another differential form β. Thus, an exact form is in the image of d, and a closed form is in the kernel of d.

For an exact form α, α = dβ for some differential form β of one-lesser degree than α. The form β is called a «potential form» or «primitive» for α. Since $d^2 = 0$, β is not unique, but can be modified by the addition of the differential of a two-step-lower-order form.

Because $d^2 = 0$, any exact form is automatically closed. The question of whether every closed form is exact depends on the topology of the domain of interest. On a contractible domain, every closed form is exact by the Poincaré lemma. More general questions of this kind on an arbitrary differentiable manifold are the subject of de Rham cohomology, which allows one to obtain purely topological information using differential methods.

Examples

The simplest example of a form which is closed but not exact is the 1-form “dθ» (quotes because it is not the derivative of a globally defined function), defined on the punctured plane $\mathbf{R}^2 \setminus \{0\}$ which is locally given as the derivative of the argument—note that argument is locally but not globally defined, since a loop around the origin increases (or decreases, depending on direction) the argument by 2π, which corresponds to the integral:

$$\oint_{S^1} d\theta = 2\pi,$$

and for general paths is known as the winding number. The differential of the argument is however globally defined (except at the origin), since differentiation only requires local data and different values of the argument differ by a constant, so the derivatives of different local definitions are equal; this line of thought is generalized in the notion of covering spaces.

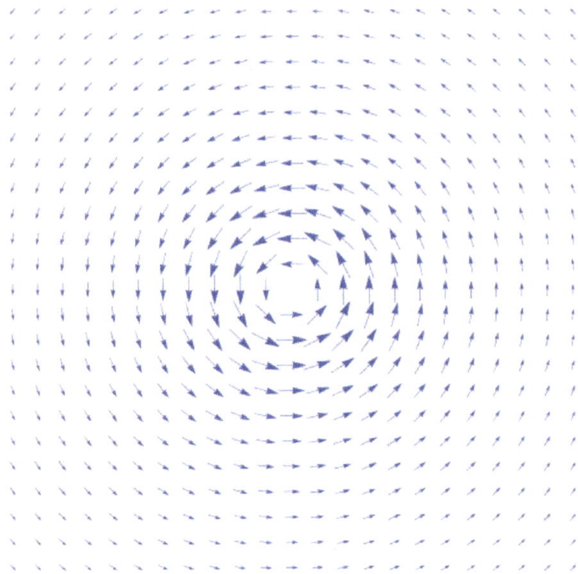

Vector field corresponding to “dθ».

Explicitly, the form is given as:

$$d\theta = \frac{1}{x^2 + y^2} \left(-y\, dx + x\, dy\right),$$

which is not defined at the origin. This can be computed from a formula for the argument, most simply via arctan(y/x) (y/x is the slope of the line passing through (x,y), and arctan converts slope to angle), recognizing $1/(x^2+y^2)$ as corresponding to the derivative of arctan, which is $1/(x^2+1)$ (these agree on the line y=1). While the differential is correctly computed by symbolically differentiating this expression, this formula is only strictly correct on the halfplane x>0, and properly one must use a correct formula for the argument.

This form generates the de Rham cohomology group $H^1_{dR}(\mathbf{R}^2 \setminus \{0\}) \cong \mathbf{R}$, meaning that any closed form ω is the sum of an exact form df and a multiple of $d\theta : \omega = df + k \cdot d\theta$, where $k = \frac{1}{2\pi} \oint_{S^1} \omega$ accounts for a non-trivial contour integral around the origin, which is the only obstruction to a closed form on the punctured plane (locally the derivative of a potential function) being the derivative of a globally defined function.

Examples in Low Dimensions

Differential forms in $\mathbf{R}^2$ and $\mathbf{R}^3$ were well known in the mathematical physics of the nineteenth century. In the plane, 0-forms are just functions, and 2-forms are functions times the basic area element dx∧dy, so that it is the 1-forms

$$\alpha = f(x,y)\,dx + g(x,y)\,dy$$

that are of real interest. The formula for the exterior derivative d here is

$$d\alpha = (g_x - f_y)\,dx \wedge dy$$

where the subscripts denote partial derivatives. Therefore the condition for α to be closed is

$$f_y = g_x.$$

In this case if h(x,y) is a function then

$$dh = h_x\,dx + h_y\,dy.$$

The implication from 'exact' to 'closed' is then a consequence of the symmetry of second derivatives, with respect to x and y.

The gradient theorem asserts that a 1-form is exact if and only if the line integral of the form depends only on the endpoints of the curve, or equivalently, if the integral around any smooth closed curve is zero.

Vector Field Analogies

On a Riemannian manifold, or more generally a pseudo-Riemannian manifold, k-forms correspond to k-vector fields (by duality via the metric), so there is a notion of a vector field corresponding to a closed or exact form.

In 3 dimensions, an exact vector field (thought of as a 1-form) is called a conservative vector field, meaning that it is the derivative (gradient) of a 0-form (function), called the scalar potential. A closed vector field (thought of as a 1-form) is one whose derivative (curl) vanishes, and is called an irrotational vector field.

Thinking of a vector field as a 2-form instead, a closed vector field is one whose derivative (divergence) vanishes, and is called an incompressible flow (sometimes solenoidal vector field).

The concepts of conservative and incompressible vector fields generalize to n dimensions, because gradient and divergence generalize to n dimensions; curl is defined only in three dimensions, thus the concept of irrotational vector field does not generalize in this way.

Poincaré Lemma

The Poincaré lemma states that if B is an open ball in R^n, any smooth closed p-form ω defined on B is exact, for any integer p with $1 \leq p \leq n$.

Translating if necessary, it can be assumed that the ball B has centre 0. Let α_s be the flow on $\mathbf{R}^n$ defined by $\alpha_s\mathbf{x} = e^{-s}\mathbf{x}$. For $s \leq 0$ it carries B into itself and induces an action on functions and differential forms. The derivative of the flow is the vector field X defined on functions f by $Xf = d(\alpha_s f)/ds|_{s=0}$: it is the radial vector field $r\partial/\partial r = \sum$

$x_i\ \partial/\partial x_i$. Often it is convenient to write the flow multiplicatively as a function of $t = e^s$, setting $\beta_t = \alpha_s$, so that $\beta_t \mathbf{x} = t\ \mathbf{x}$. Only if $0 < \mathbf{t} \leq 1$ will β_t carry B into itself. The derivative of the flow on forms defines the Lie derivative with respect to X given by $L_X \omega = d(\alpha_s \omega)\ /ds|_{s=0}$. In particular

$$\frac{d}{ds}\alpha_s \omega = \alpha_s L_X \omega,$$

so by the chain rule

$$\frac{d}{dt}\beta_t \omega = t^{-1}\beta_t L_X \omega.$$

Since α_s commutes with the exterior derivative d, so does the Lie derivative L_X. Moreover if ι_X denotes interior multiplication or contraction by the vector field X, then by Cartan's formula

$$d\iota_X + \iota_X d = L_X.$$

Now define

$$h\omega = \int_0^1 \beta_t \omega \frac{dt}{t}.$$

Then h commutes with L_X, since L_X commutes with α_s and β_t. Furthermore h commutes with d. It also has the important property that

$$h \circ L_X = \mathrm{id} = L_X \circ h$$

by the fundamental theorem of calculus since

$$hL_X \omega = \int_0^1 \beta_t L_X \omega \frac{dt}{t} = \int_0^1 \frac{d}{dt}(\beta_t \omega)\, dt = [\beta_t \omega]_0^1 = \omega.$$

Now set $k = h \circ \iota_X$. Then

$$dk + kd = dh\iota_X + h\iota_X d = h(d\iota_X + \iota_X d) = hL_X = \mathrm{id}.$$

Thus

$$dk + kd = \mathrm{id}.$$

(In the language of homological algebra, k is a "contracting homotopy".)

It now follows that if ω is closed, so that $d\omega = 0$, then $d(k\omega) = (dk + kd)\omega = \omega$, so that ω is exact and the Poincaré lemma is proved.

The same method applies to any open set in $\mathbf{R}^n$ that is star-shaped about 0, i.e. any open set containing 0 and invariant under β_t for $0 < t < 1$.

Example. In two dimensions the Poincaré lemma can be proved directly for closed 1-forms and 2-forms as follows.

If $\omega = p\,dx + q\,dy$ is a closed 1-form on $(a,b) \times (c,d)$, then $p_y = q_x$. If $\omega = df$ then $p = f_x$ and $q = f_y$. Set

$$g(x,y) = \int_a^x p(t,y)\,dt,$$

so that $g_x = p$. Then $h = f - g$ must satisfy $h_x = 0$ and $h_y = q - g_y$. The right hand side here is independent of x since its partial derivative with respect to x is 0. So

$$h(x,y) = \int_c^y q(a,s)\,ds - g(a,y) = \int_c^y q(a,s)\,ds,$$

and hence

$$f(x,y) = \int_a^x p(t,y)\,dt + \int_c^y q(a,s)\,ds.$$

Similarly if $\Omega = r\,dx \wedge dy$ then $\Omega = d(a\,dx + b\,dy)$ with $b_x - a_y = r$. Thus a solution is given by $a = 0$ and

$$b(x,y) = \int_a^x r(t,y)\,dt.$$

Formulation as Cohomology

When the difference of two closed forms is an exact form, they are said to be cohomologous to each other. That is, if ζ and η are closed forms, and one can find some β such that

$$\zeta - \eta = d\beta$$

then one says that ζ and η are cohomologous to each other. Exact forms are sometimes said to be cohomologous to zero. The set of all forms cohomologous to a given form (and thus to each other) is called a de Rham cohomology class; the general study of such classes is known as cohomology. It makes no real sense to ask whether a 0-form (smooth function) is exact, since d increases degree by 1; but the clues from topology suggest that only the zero function should be called "exact". The cohomology classes are identified with locally constant functions.

Using contracting homotopies similar to the one used in the proof of the Poincaré lemma, it can be shown that de Rham cohomology is homotopy-invariant. Non-contractible in general have non-trivial de Rham cohomology. For instance, on the circle S^1, parametrized by t in [0, 1], the closed 1-form dt is not exact.

Application in Electrodynamics

In electrodynamics, the case of the magnetic field $\vec{B}(\mathbf{r})$ produced by a stationary electrical current is important. There one deals with the vector potential $\vec{A}(\mathbf{r})$ of this field. This case corresponds to k=2, and the defining region is the full $\mathbb{R}^3$ The current-density vector is $\vec{j}$. It corresponds to the current two-form

$$\mathbf{I} := j_1(x,y,z)\,\mathrm{d}x_2 \wedge \mathrm{d}x_3 + j_2(x,y,z)\,\mathrm{d}x_3 \wedge \mathrm{d}x_1 + j_3(x,y,z)\,\mathrm{d}x_1 \wedge \mathrm{d}x_2.$$

For the magnetic field $\vec{B}$one has analogous results: it corresponds to the induction two-form $\Phi_B := B_1 \mathrm{d}x_2 \wedge \mathrm{d}x_3 + \cdots$ and can be derived from the vector potential $\vec{A}$, or the corresponding one-form A,

$$\vec{B} = \operatorname{curl} \vec{A} = \left\{ \frac{\partial A_3}{\partial x_2} - \frac{\partial A_2}{\partial x_3}, \frac{\partial A_1}{\partial x_3} - \frac{\partial A_3}{\partial x_1}, \frac{\partial A_2}{\partial x_1} - \frac{\partial A_1}{\partial x_2} \right\}, \text{ or } \Phi_B = \mathrm{d}\mathbf{A}.$$

Thereby the vector potential $\vec{A}$ corresponds to the potential one-form

$$\mathbf{A} := A_1\,\mathrm{d}x_1 + A_2\,\mathrm{d}x_2 + A_3\,\mathrm{d}x_3.$$

The closedness of the magnetic-induction two-form corresponds to the property of the magnetic field that it is source-free: $\operatorname{div} \vec{B} \equiv 0$,i.e. there are no magnetic monopoles.

In a special gauge, $\operatorname{div} \vec{A} \stackrel{!}{=} 0$, this implies for i = 1, 2, 3

$$A_i(\vec{r}) = \int \frac{\mu_0 j_i(\vec{r}\,')\; dx_1' dx_2' dx_3'}{4\pi|\vec{r} - \vec{r}\,'|}.$$

(Here μ_0 is a constant, the magnetic vacuum permeability.)

This equation is remarkable, because it corresponds completely to a well-known formula for the electrical field $\vec{E}$, namely for the electrostatic Coulomb potential $\phi(x_1, x_2, x_3)$ of a charge density $\rho(x_1, x_2, x_3)$. At this place one can already guess that

- $\vec{E}$ and $\vec{B}$,
- ρ and $\vec{j}$,
- ϕ and $\vec{A}$

can be unified to quantities with six rsp. four nontrivial components, which is the basis of the relativistic invariance of the Maxwell equations.

If the condition of stationarity is left, on the l.h.s. of the above-mentioned equation one must add, in the equations for $A_i$to the three space coordinates, as a fourth variable also the time t, whereas on the r.h.s., in j'_i the so-called "retarded time", $t' := t - \frac{|\vec{r} - \vec{r}\,'|}{c}$ must be used, i.e. it is added to the argument of the current-density. Finally, as before, one integrates over the three primed space coordinates. (As usual c is the vacuum velocity of light.)

REFERENCES

1. Newton, Isaac. (c.1671). Methodus Fluxionum et Serierum Infinitarum (The Method of Fluxions and Infinite Series), published in 1736 [Opuscula, 1744, Vol. I. p. 66].
2. Bernoulli, Jacob (1695), "Explicationes, Annotationes & Additiones ad ea, quae in Actis sup. de Curva Elastica, Isochrona Paracentrica, & Velaria, hinc inde memorata, & paratim controversa legundur; ubi de Linea mediarum directionum, alliisque novis", Acta Eruditorum
3. Hairer, Ernst; Nørsett, Syvert Paul; Wanner, Gerhard (1993), Solving ordinary differential equations I: Nonstiff problems, Berlin, New York: Springer-Verlag, ISBN 978-3-540-56670-0
4. Cannon, John T.; Dostrovsky, Sigalia (1981). "The evolution of dynamics, vibration theory from 1687 to 1742". Studies in the History of Mathematics and Physical Sciences. 6. New York: Springer-Verlag: ix + 184 pp. ISBN 0-3879-0626-6. GRAY, JW (July 1983). "BOOK REVIEWS". BULLETIN (New Series) OF THE AMERICAN MATHEMATICAL SOCIETY. 9 (1). (retrieved 13 Nov 2012).
5. Wheeler, Gerard F.; Crummett, William P. (1987). "The Vibrating String Controversy". Am. J. Phys. 55 (1): 33–37. Bibcode:1987AmJPh..55...33W. doi:10.1119/1.15311.
6. For a special collection of the 9 groundbreaking papers by the three authors, see First Appearance of the wave equation: D'Alembert, Leonhard Euler, Daniel Bernoulli. - the controversy about vibrating strings (retrieved 13 Nov 2012). Herman HJ Lynge and Son.
7. For de Lagrange's contributions to the acoustic wave equation, can consult Acoustics: An Introduction to Its Physical Principles and Applications Allan D. Pierce, Acoustical Soc of America, 1989; page 18.(retrieved 9 Dec 2012)
8. Speiser, David. Discovering the Principles of Mechanics 1600-1800, p. 191 (Basel: Birkhäuser, 2008).
9. Fourier, Joseph (1822). Théorie analytique de la chaleur (in French). Paris: Firmin Didot Père et Fils. OCLC 2688081.
10. Boyce, William E.; DiPrima, Richard C. (1967). Elementary Differential Equations and Boundary Value Problems (4th ed.). John Wiley & Sons. p. 3.
11. Weisstein, Eric W. "Ordinary Differential Equation Order." From MathWorld--A Wolfram Web Resource. http://mathworld.wolfram.com/OrdinaryDifferentialEquationOrder.html

12. Order and degree of a differential equation, accessed Dec 2015.
13. Zill, Dennis G. A First Course in Differential Equations (5th ed.). Brooks/Cole. ISBN 0-534-37388-7.
14. Einstein, Albert (1916). "The Foundation of the General Theory of Relativity". Annalen der Physik. 354 (7): 769. Bibcode:1916AnP...354..769E. doi:10.1002/andp.19163540702. Archived from the original (PDF) on 2006-08-29.
15. Einstein, Albert (November 25, 1915). "Die Feldgleichungen der Gravitation". Sitzungsberichte der Preussischen Akademie der Wissenschaften zu Berlin: 844–847. Retrieved 2006-09-12.
16. Misner, Charles W.; Thorne, Kip S.; Wheeler, John Archibald (1973). Gravitation. San Francisco: W. H. Freeman. ISBN 978-0-7167-0344-0 Chapter 34, p. 916.
17. Griffiths, David J. (2004), Introduction to Quantum Mechanics (2nd ed.), Prentice Hall, pp. 1–2, ISBN 0-13-111892-7
18. IUPAC Gold Book definition of rate law. See also: According to IUPAC Compendium of Chemical Terminology.
19. Kenneth A. Connors Chemical Kinetics, the study of reaction rates in solution, 1991, VCH Publishers.
20. Eremenko, A. (1982). "Meromorphic solutions of algebraic differential equations" (PDF). Russian mathematical Surveys. 37: 61–94.
21. So-Chin Chen; Mei-Chi Shaw (2002). Partial Differential Equations in Several Complex Variables. American Mathematical Society. ISBN 978-0-8218-2961-5.
22. Complex Difference Equations of Malmquist Type
23. An Introduction to complex functions on product of two time scales
24. Analytic solutions to integral equations in the complex domain

CHAPTER 2

Ordinary Differential Equations

Chapter Contents

In mathematics, an ordinary differential equation (ODE) is a differential equation containing one or more functions of one independent variable and its derivatives. The term ordinary is used in contrast with the term partial differential equation which may be with respect to more than one independent variable.

ODEs that are linear differential equations have exact closed-form solutions that can be added and multiplied by coefficients. By contrast, ODEs that lack additive solutions are nonlinear, and solving them is far more intricate, as one can rarely represent them by elementary functions in closed form: Instead, exact and analytic solutions of ODEs are in series or integral form. Graphical and numerical methods, applied by hand or by computer, may approximate solutions of ODEs and perhaps yield useful information, often sufficing in the absence of exact, analytic solutions.

INTRODUCTION

Ordinary differential equations (ODEs) arise in many contexts of mathematics and science (social as well as natural). Mathematical descriptions of change use differentials and derivatives. Various differentials, derivatives, and functions become related to each other via equations, and thus a differential equation is a result that describes dynamically changing phenomena, evolution, and variation. Often, quantities are defined as the rate of change of other quantities (for example, derivatives of displacement with respect to time), or gradients of quantities, which is how they enter differential equations.

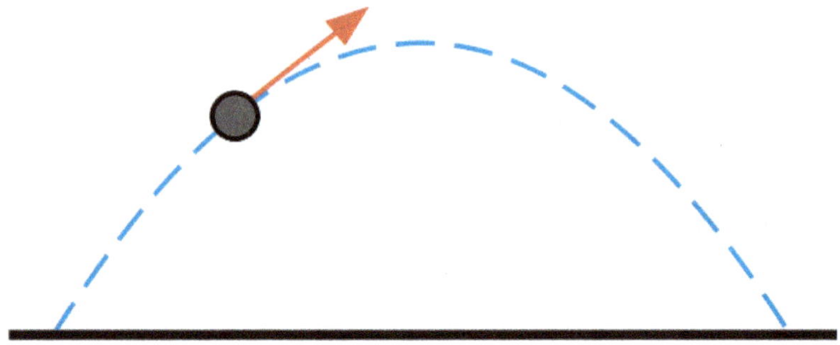

The trajectory of a projectile launched from a cannon follows a curve determined by an ordinary differential equation that is derived from Newton's second law.

Specific mathematical fields include geometry and analytical mechanics. Scientific fields include much of physics and astronomy (celestial mechanics), meteorology (weather modelling), chemistry (reaction rates), biology (infectious diseases, genetic variation), ecology and population modelling (population competition), economics (stock trends, interest rates and the market equilibrium price changes).

Many mathematicians have studied differential equations and contributed to the field, including Newton, Leibniz, the Bernoulli family, Riccati, Clairaut, d'Alembert, and Euler.

A simple example is Newton's second law of motion — the relationship between the displacement x and the time t of an object under the force F, is given by the differential equation

$$m\frac{\mathrm{d}^2 x(t)}{\mathrm{d}t^2} = F(x(t))$$

which constrains the motion of a particle of constant mass m. In general, F is a function of the position x(t) of the particle at time t. The unknown function x(t) appears on both sides of the differential equation, and is indicated in the notation F(x(t)).

DEFINITIONS

In what follows, let y be a dependent variable and x an independent variable, and y = f(x) is an unknown function of x. The notation for differentiation varies depending upon the author and upon which notation is most useful for the task at hand. In this context, the Leibniz's notation ($dy/dx, d^2y/dx^2, \ldots d^ny/dx^n$) is more useful for differentiation and integration, whereas Newton's and Lagrange's notation (y', y'', ... $y^{(n)}$) is more useful for representing derivatives of any order compactly.

General Definition of an ODE

Given F, a function of x, y, and derivatives of y. Then an equation of the form

$$F\left(x, y, y', \cdots y^{(n-1)}\right) = y^{(n)}$$

is called an explicit ordinary differential equation of order n.

More generally, an implicit ordinary differential equation of order n takes the form:

$$F\left(x, y, y', y'', \cdots, y^{(n)}\right) = 0$$

There are further classifications:

Autonomous

A differential equation not depending on x is called autonomous.

Linear

A differential equation is said to be linear if F can be written as a linear combination of the derivatives of y:

$$y^{(n)} = \sum_{i=0}^{n-1} a_i(x) y^{(i)} + r(x)$$

where $a_i(x)$ and r(x) are continuous functions in x. The function r(x) is called the source term, leading to two further important classifications:

Homogeneous

If r(x) = 0, and consequently one "automatic" solution is the trivial solution, y = 0. The solution of a linear homogeneous equation is a complementary function, denoted here by y_c.

Nonhomogeneous (or inhomogeneous)

If r(x) ≠ 0. The additional solution to the complementary function is the **particular integral**, denoted here by y_p.

The general solution to a linear equation can be written as $y = y_c + y_p$.

Non-linear

A differential equation that cannot be written in the form of a linear combination.

System of ODEs

A number of coupled differential equations form a system of equations. If **y** is a vector whose elements are functions; **y**(x) = [y_1(x), y_2(x),..., y_m(x)], and **F** is a vector-valued function of **y** and its derivatives, then

$$\mathbf{y}^{(n)} = \mathbf{F}\left(x, \mathbf{y}, \mathbf{y}', \mathbf{y}'', \cdots \mathbf{y}^{(n-1)}\right)$$

is an explicit system of ordinary differential equations of order or dimension m. In column vector form:

$$\begin{pmatrix} y_1^{(n)} \\ y_2^{(n)} \\ \vdots \\ y_m^{(n)} \end{pmatrix} = \begin{pmatrix} f_1\left(x, \mathbf{y}, \mathbf{y}', \mathbf{y}'', \cdots \mathbf{y}^{(n-1)}\right) \\ f_2\left(x, \mathbf{y}, \mathbf{y}', \mathbf{y}'', \cdots \mathbf{y}^{(n-1)}\right) \\ \vdots \\ f_m\left(x, \mathbf{y}, \mathbf{y}', \mathbf{y}'', \cdots \mathbf{y}^{(n-1)}\right) \end{pmatrix}$$

These are not necessarily linear. The implicit analogue is:

$$\mathbf{F}\left(x, \mathbf{y}, \mathbf{y}', \mathbf{y}'', \cdots \mathbf{y}^{(n)}\right) = \mathbf{0}$$

where **0** = (0, 0,... 0) is the zero vector. In matrix form

$$\begin{pmatrix} f_1(x, \mathbf{y}, \mathbf{y}', \mathbf{y}'', \cdots \mathbf{y}^{(n)}) \\ f_2(x, \mathbf{y}, \mathbf{y}', \mathbf{y}'', \cdots \mathbf{y}^{(n)}) \\ \vdots \\ f_m(x, \mathbf{y}, \mathbf{y}', \mathbf{y}'', \cdots \mathbf{y}^{(n)}) \end{pmatrix} = \begin{pmatrix} 0 \\ 0 \\ \vdots \\ 0 \end{pmatrix}$$

For a system of the form $\mathbf{F}\left(x, \mathbf{y}, \mathbf{y}'\right) = \mathbf{0}$, some sources also require that the Jacobian matrix $\frac{\partial \mathbf{F}(x, \mathbf{u}, \mathbf{v})}{\partial \mathbf{v}}$ be non-singular in order to call this an implicit ODE [system]; an implicit ODE system satisfying this Jacobian non-singularity condition can be transformed into an explicit ODE system. In the same sources, implicit ODE systems with a singular Jacobian are termed differential algebraic equations (DAEs). This distinction is not merely one of terminology; DAEs have fundamentally different characteristics and are generally more involved to solve than (nonsingular) ODE systems. Presumably for additional derivatives, the Hessian matrix and so forth are also assumed non-singular according to this scheme, although note that any ODE of order greater than one can be [and usually is] rewritten as system of ODEs of first order, which makes the Jacobian singularity criterion sufficient for this taxonomy to be comprehensive at all orders.

Solutions

Given a differential equation

$$F\left(x, y, y', \cdots, y^{(n)}\right) = 0$$

a function u: I ⊂ **R** → **R** is called the solution or integral curve for F, if u is n-times differentiable on I, and

$$F(x, u, u', \cdots, u^{(n)}) = 0 \quad x \in I.$$

Given two solutions u: J ⊂ **R** → **R** and v: I ⊂ **R** → **R**, u is called an extension of v if I ⊂ J and

$$u(x) = v(x) \quad x \in I.$$

A solution that has no extension is called a maximal solution. A solution defined on all of **R** is called a global solution.

A general solution of an nth-order equation is a solution containing n arbitrary independent constants of integration. A particular solution is derived from the general solution by setting the constants to particular values, often chosen to fulfill set 'initial conditions or boundary conditions'. A singular solution is a solution that cannot be obtained by assigning definite values to the arbitrary constants in the general solution.

THEORIES OF ODES

Singular Solutions

The theory of singular solutions of ordinary and partial differential equations was a subject of research from the time of Leibniz, but only since the middle of the nineteenth century did it receive special attention. A valuable but little-known work on the subject is that of Houtain (1854). Darboux (starting in 1873) was a leader in the theory, and in the geometric interpretation of these solutions he opened a field worked by various writers, notable ones being Casorati and Cayley. To the latter is due (1872) the theory of singular solutions of differential equations of the first order as accepted circa 1900.

Reduction to Quadratures

The primitive attempt in dealing with differential equations had in view a reduction to quadratures. As it had been the hope of eighteenth-century algebraists to find a method for solving the general equation of the nth degree, so it was the hope of analysts to find a general method for integrating any differential equation. Gauss (1799) showed, however, that the differential equation meets its limitations very soon unless complex numbers are introduced. Hence, analysts began to substitute the study of functions, thus opening a new and fertile field. Cauchy was the first to appreciate the importance of this view. Thereafter, the real question was to be not whether a solution is possible by means of known functions or their integrals but whether a given differential equation suffices for the definition of a function of the independent variable or variables, and, if so, what are the characteristic properties of this function.

Fuchsian Theory

Two memoirs by Fuchs (Crelle, 1866, 1868), inspired a novel approach, subsequently elaborated by Thomé and Frobenius. Collet was a prominent contributor beginning in 1869, although his method for integrating a non-linear system was communicated to Bertrand in 1868. Clebsch (1873) attacked the theory along lines parallel to those followed in his theory of Abelian integrals. As the latter can be classified according to the properties of the fundamental curve that remains unchanged under a rational transformation, so Clebsch proposed to classify the transcendent functions defined by the differential equations according to the invariant properties of the corresponding surfaces f = 0 under rational one-to-one transformations.

Frobenius Method

In mathematics, the Method of Frobenius, named after Ferdinand Georg Frobenius, is a way to find an infinite series solution for a second-order ordinary differential equation of the form

$$z^2u'' + p(z)zu' + q(z)u = 0$$

with

$$u' \equiv \frac{du}{dz} \quad \text{and} \quad u'' \equiv \frac{d^2u}{dz^2}$$

in the vicinity of the regular singular point z=0. We can divide by z^2 **to obtain a differential equation of the form**

$$u'' + \frac{p(z)}{z}u' + \frac{q(z)}{z^2}u = 0$$

which will not be solvable with regular power series methods if either p(z)/z **or** q(z)/z^2 **are not analytic at** z = **0. The** Frobenius method enables us to create a power series solution to such a differential equation, provided that p(z) and q(z) are themselves analytic at 0 or, being analytic elsewhere, both their limits at 0 exist (and are finite).

Explanation

The Method of Frobenius tells us that we can seek a power series solution of the form

$$u(z) = \sum_{k=0}^{\infty} A_k z^{k+r}, \qquad (A_0 \neq 0)$$

Differentiating:

$$u'(z) = \sum_{k=0}^{\infty}(k+r)A_k z^{k+r-1}$$

$$u''(z) = \sum_{k=0}^{\infty}(k+r-1)(k+r)A_k z^{k+r-2}$$

Substituting:

$$z^2\sum_{k=0}^{\infty}(k+r-1)(k+r)A_k z^{k+r-2} + zp(z)\sum_{k=0}^{\infty}(k+r)A_k z^{k+r-1} + q(z)\sum_{k=0}^{\infty}A_k z^{k+r}$$

$$= \sum_{k=0}^{\infty}(k+r-1)(k+r)A_k z^{k+r} + p(z)\sum_{k=0}^{\infty}(k+r)A_k z^{k+r} + q(z)\sum_{k=0}^{\infty}A_k z^{k+r}$$

$$= \sum_{k=0}^{\infty}[(k+r-1)(k+r)A_k z^{k+r} + p(z)(k+r)A_k z^{k+r} + q(z)A_k z^{k+r}]$$

$$= \sum_{k=0}^{\infty}[(k+r-1)(k+r) + p(z)(k+r) + q(z)]\,A_k z^{k+r}$$

$$= [r(r-1) + p(z)r + q(z)]\,A_0 z^r + \sum_{k=1}^{\infty}[(k+r-1)(k+r) + p(z)(k+r) + q(z)]\,A_k z^{k+r}$$

The expression

$$r\,(r-1) + p\,(0)\,r + q\,(0) = I(r)$$

is known as the indicial polynomial, which is quadratic in r. The general definition of the indicial polynomial is the coefficient of the lowest power of z in the infinite series. In this case it happens to be that this is the rth coefficient but, it is possible for the lowest possible exponent to be r − 2, r − 1 or, something else depending on the given differential equation. This detail is important to keep in mind. In the process of synchronizing all the series of the differential equation to start at the same index value (which in the above expression is k = 1), one can end up with complicated expressions. However, in solving for the indicial roots attention is focused only on the coefficient of the lowest power of z.

Using this, the general expression of the coefficient of z^{k+r} **is**

$$I(k+r)A_k + \sum_{j=0}^{k-1}\frac{(j+r)p^{(k-j)}(0) + q^{(k-j)}(0)}{(k-j)!}A_j,$$

These coefficients must be zero, since they should be solutions of the differential equation, so

$$I(k+r)A_k + \sum_{j=0}^{k-1} \frac{(j+r)p^{(k-j)}(0) + q^{(k-j)}(0)}{(k-j)!} A_j = 0$$

$$\sum_{j=0}^{k-1} \frac{(j+r)p^{(k-j)}(0) + q^{(k-j)}(0)}{(k-j)!} A_j = -I(k+r)A_k$$

$$\frac{1}{-I(k+r)} \sum_{j=0}^{k-1} \frac{(j+r)p^{(k-j)}(0) + q^{(k-j)}(0)}{(k-j)!} A_j = A_k$$

The series solution with A_k **above,**

$$U_r(z) = \sum_{k=0}^{\infty} A_k z^{k+r}$$

satisfies

$$z^2 U_r(z)'' + p(z) z U_r(z)' + q(z) U_r(z) = I(r) z^r$$

If we choose one of the roots to the indicial polynomial for r **in** $U_r(z)$, we gain a solution to the differential equation. If the difference between the roots is not an integer, we get another, linearly independent solution in the other root.

Example

Let us solve

$$z^2 f'' - z f' + (1-z) f = 0$$

Divide throughout by z^2 to give

$$f'' - \frac{1}{z} f' + \frac{1-z}{z^2} f = f'' - \frac{1}{z} f' + \left(\frac{1}{z^2} - \frac{1}{z}\right) f = 0$$

which has the requisite singularity at z = **0.**

Use the series solution

$$f = \sum_{k=0}^{\infty} A_k z^{k+r}$$

$$f' = \sum_{k=0}^{\infty} (k+r) A_k z^{k+r-1}$$

$$f'' = \sum_{k=0}^{\infty} (k+r)(k+r-1) A_k z^{k+r-2}$$

Now, substituting

$$\sum_{k=0}^{\infty}(k+r)(k+r-1)A_k z^{k+r-2} - \frac{1}{z}\sum_{k=0}^{\infty}(k+r)A_k z^{k+r-1} + \left(\frac{1}{z^2}-\frac{1}{z}\right)\sum_{k=0}^{\infty}A_k z^{k+r}$$

$$= \sum_{k=0}^{\infty}(k+r)(k+r-1)A_k z^{k+r-2} - \frac{1}{z}\sum_{k=0}^{\infty}(k+r)A_k z^{k+r-1} + \frac{1}{z^2}\sum_{k=0}^{\infty}A_k z^{k+r} - \frac{1}{z}\sum_{k=0}^{\infty}A_k z^{k+r}$$

$$= \sum_{k=0}^{\infty}(k+r)(k+r-1)A_k z^{k+r-2} - \sum_{k=0}^{\infty}(k+r)A_k z^{k+r-2} + \sum_{k=0}^{\infty}A_k z^{k+r-2} - \sum_{k=0}^{\infty}A_k z^{k+r-1}$$

$$= \sum_{k=0}^{\infty}(k+r)(k+r-1)A_k z^{k+r-2} - \sum_{k=0}^{\infty}(k+r)A_k z^{k+r-2} + \sum_{k=0}^{\infty}A_k z^{k+r-2} - \sum_{k-1=0}^{\infty}A_{k-1} z^{k-1+r-1}$$

$$= \sum_{k=0}^{\infty}(k+r)(k+r-1)A_k z^{k+r-2} - \sum_{k=0}^{\infty}(k+r)A_k z^{k+r-2} + \sum_{k=0}^{\infty}A_k z^{k+r-2} - \sum_{k=1}^{\infty}A_{k-1} z^{k+r-2}$$

$$= \left\{\sum_{k=0}^{\infty}\left((k+r)(k+r-1)-(k+r)+1\right)A_k z^{k+r-2}\right\} - \sum_{k=1}^{\infty}A_{k-1} z^{k+r-2}$$

$$= \left\{\left(r(r-1)-r+1\right)A_0 z^{r-2} + \sum_{k=1}^{\infty}\left((k+r)(k+r-1)-(k+r)+1\right)A_k z^{k+r-2}\right\} - \sum_{k=1}^{\infty}A_{k-1} z^{k+r-2}$$

$$= (r-1)^2 A_0 z^{r-2} + \left\{\sum_{k=1}^{\infty}(k+r-1)^2 A_k z^{k+r-2} - \sum_{k=1}^{\infty}A_{k-1} z^{k+r-2}\right\}$$

$$= (r-1)^2 A_0 z^{r-2} + \sum_{k=1}^{\infty}\left((k+r-1)^2 A_k - A_{k-1}\right) z^{k+r-2}$$

From $(r - 1)^2 = 0$ we get a double root of 1. Using this root, we set the coefficient of z^{k+r-2} to be zero (for it to be a solution), which gives us:

$$(k+1-1)^2 A_k - A_{k-1} = k^2 A_k - A_{k-1} = 0$$

hence we have the recurrence relation:

$$A_k = \frac{A_{k-1}}{k^2}$$

Given some initial conditions, we can either solve the recurrence entirely or obtain a solution in power series form.

Since the ratio of coefficients A_k/A_{k-1} **is a rational function, the power series can be written as a generalized** hypergeometric series.

Roots Separated by an Integer

The previous example involved an indicial polynomial with a repeated root, which gives only one solution to the given differential equation. In general, the Frobenius method gives two independent solutions provided that the indicial equation's roots are not separated by an integer (including zero).

If the root is repeated or the roots differ by an integer, then the second solution can be found using:

$$y_2 = Cy_1 \ln x + \sum_{k=0}^{\infty} B_k x^{k+r_2}$$

where $y_1(x)$ is the first solution (based on the larger root in the case of unequal roots), r_2 is the smaller root, and the constant C and the coefficients B_k are to be determined. Once B_0 is chosen (for example by setting it to 1) then C and the B_k are determined up to but not including $B_{r_1-r_2}$, which can be set arbitrarily. This then determines the rest of the B_k In some cases the constant C must be zero. For example, consider the following differential equation (Kummer's equation with a = 1 and b = 2):

$$zu'' + (2 - z)u' - u = 0$$

The roots of the indicial equation are −1 and 0. Two independent solutions are $1/z$ and $(e^z)/z$ so we see that the logarithm does not appear in any solution. The solution $(e^z - 1)/z$ has a power series starting with the power zero. In a power series starting with z^{-1} the recurrence relation places no restriction on the coefficient for the term z^0 which can be set arbitrarily. If it is set to zero then with this differential equation all the other coefficients will be zero and we obtain the solution 1/z.

Lie's Theory

From 1870, Sophus Lie's work put the theory of differential equations on a more satisfactory foundation. He showed that the integration theories of the older mathematicians can, by the introduction of what are now called Lie groups, be referred to a common source, and that ordinary differential equations that admit the same infinitesimal transformations present comparable difficulties of integration. He also emphasized the subject of transformations of contact. Lie's group theory of differential equations has been certified, namely: (1) that it unifies the many ad hoc methods known for solving differential equations, and (2) that it provides powerful new ways to find solutions. The theory has applications to both ordinary and partial differential equations.

A general approach to solve DEs uses the symmetry property of differential equations, the continuous infinitesimal transformations of solutions to solutions (Lie theory). Continuous group theory, Lie algebras, and differential geometry are used to understand the structure of linear and nonlinear (partial) differential equations for generating integrable equations, to find its Lax pairs, recursion operators, Bäcklund transform, and finally finding exact analytic solutions to the DE.

Symmetry methods have been recognized to study differential equations, arising in mathematics, physics, engineering, and many other disciplines.

Sturm–Liouville Theory

In mathematics and its applications, a classical Sturm–Liouville theory, named after Jacques Charles François Sturm (1803–1855) and Joseph Liouville (1809–1882), is the theory of a real second-order linear differential equation of the form

$$\frac{d}{dx}\left[p(x)\frac{dy}{dx}\right]+q(x)y=-\lambda w(x)y,$$

where y is a function of the free variable x. Here the functions p(x), q(x), and w(x) > 0 are specified at the outset. In the simplest of cases all coefficients are continuous on the finite closed interval [a,b], and p has continuous derivative. In this simplest of all cases, this function "y" is called a solution if it is continuously differentiable on (a,b) and satisfies the equation **('1')** at every point in (a,b). In addition, the unknown function y is typically required to satisfy some boundary conditions at a and b. The function w(x), which is sometimes called r(x), is called the "weight" or "density" function.

The value of λ is not specified in the equation; finding the values of λ for which there exists a non-trivial solution of **('1')** satisfying the boundary conditions is part of the problem called the **Sturm–Liouville (S–L) problem**.

Such values of λ, when they exist, are called the eigenvalues of the boundary value problem defined by **('1')** and the prescribed set of boundary conditions. The corresponding solutions (for such a λ) are the eigenfunctions of this problem. Under normal assumptions on the coefficient functions p(x), q(x), and w(x) above, they induce a Hermitian differential operator in some function space defined by boundary conditions. The resulting theory of the existence and asymptotic behavior of the eigenvalues, the corresponding qualitative theory of the eigenfunctions and their completeness in a suitable function space became known as **Sturm–Liouville theory**. This theory is important in applied mathematics, where S–L problems occur very commonly, particularly when dealing with linear partial differential equations that are separable.

A Sturm–Liouville (S–L) problem is said to be regular if p(x), w(x) > 0, and p(x), p'(x), q(x), and w(x) are continuous functions over the finite interval [a, b], and has separated boundary conditions of the form

$$\alpha_1 y(a)+\alpha_2 y'(a)=0 \qquad (\alpha_1^2+\alpha_2^2>0),$$

$$\beta_1 y(b)+\beta_2 y'(b)=0 \qquad (\beta_1^2+\beta_2^2>0),$$

Under the assumption that the S–L problem is regular, the main tenet of **Sturm–Liouville theory** states that:

- The eigenvalues λ_1, λ_2, λ_3, ... of the regular Sturm–Liouville problem **('1')-('2')-('3')** are real and can be ordered such that

$$\lambda_1<\lambda_2<\lambda_3<\cdots<\lambda_n<\cdots\to\infty;$$

- Corresponding to each eigenvalue λ_n is a unique (up to a normalization constant) eigenfunction $y_n(x)$ which has exactly n − 1 zeros in (a, b). The eigenfunction $y_n(x)$ is called the n-th fundamental solution satisfying the regular Sturm–Liouville problem **('1')-('2')-('3')**.
- The normalized eigenfunctions form an orthonormal basis

$$\int_a^b y_n(x)y_m(x)w(x)\,\mathrm{d}x = \delta_{mn},$$

in the Hilbert space $L^2([a, b], w(x)dx)$. Here δ_{mn} is the Kronecker delta.

Note that, unless p(x) is continuously differentiable and q(x), w(x) are continuous, the equation has to be understood in a weak sense.

Sturm–Liouville Form

The differential equation **('1')** is said to be in **Sturm–Liouville form** or **self-adjoint form**. All second-order linear ordinary differential equations can be recast in the form on the left-hand side of **('1')** by multiplying both sides of the equation by an appropriate integrating factor (although the same is not true of second-order partial differential equations, or if y is a vector.)

Examples

The Bessel Equation]

$$x^2y'' + xy' + (x^2 - \nu^2)\,y = 0$$

which can be written in Sturm–Liouville form as

$$(xy')' + \left(x - \frac{\nu^2}{x}\right) y = 0.$$

The Legendre Equation

$$(1 - x^2)y'' - 2xy' + \nu(\nu + 1)y = 0$$

which can easily be put into Sturm–Liouville form, since $\frac{d}{dx}(1 - x^2) = -2x$, so the Legendre equation is equivalent to

$$[(1 - x^2)y']' + \nu(\nu + 1)y = 0$$

An Example Using An Integrating Factor

$$x^3y'' - xy' + 2y = 0.$$

Divide throughout by x^3:

$$y'' - \frac{1}{x^2}y' + \frac{2}{x^3}y = 0$$

Multiplying throughout by an integrating factor of

$$\mu(x) = e^{\int -\frac{1}{x^2}\,\mathrm{d}x} = e^{\frac{1}{x}},$$

gives

$$e^{\frac{1}{x}}y'' - \frac{e^{\frac{1}{x}}}{x^2}y' + \frac{2e^{\frac{1}{x}}}{x^3}y = 0$$

which can be easily put into Sturm–Liouville form since

$$\frac{d}{dx}e^{\frac{1}{x}} = -\frac{e^{\frac{1}{x}}}{x^2}$$

so the differential equation is equivalent to

$$(e^{\frac{1}{x}}y')' + \frac{2e^{\frac{1}{x}}}{x^3}y = 0.$$

The Integrating Factor For A General Second Order Differential Equation]

$P(x)y'' + Q(x)y' + R(x)y = 0$

multiplying through by the integrating factor

$$\mu(x) = \frac{1}{P(x)}e^{\int \frac{Q(x)}{P(x)}\mathrm{d}x},$$

and then collecting gives the Sturm–Liouville form:

$$\frac{d}{dx}(\mu(x)P(x)y') + \mu(x)R(x)y = 0$$

or, explicitly,

$$\frac{d}{dx}\left(e^{\int \frac{Q(x)}{P(x)}\mathrm{d}x}y'\right) + \frac{R(x)}{P(x)}e^{\int \frac{Q(x)}{P(x)}\,\mathrm{d}x}y = 0$$

Sturm–Liouville Equations as self-Adjoint Differential Operators]

The map

$$Lu = -\frac{1}{w(x)}\left(\frac{\mathrm{d}}{\mathrm{d}x}\left[p(x)\frac{\mathrm{d}u}{\mathrm{d}x}\right] + q(x)u\right)$$

can be viewed as a linear operator mapping a function u to another function Lu. One may study this linear operator in the context of functional analysis. In fact, equation ('1') can be written as

$Lu = \lambda u.$

This is precisely the eigenvalue problem; that is, one is trying to find the eigenvalues $\lambda_1, \lambda_2, \lambda_3, \ldots$ and the corresponding eigenvectors $u_1, u_2, u_3, \ldots$ of the L operator. The proper setting for this problem is the Hilbert space $L^2([a, b], w(x)\,dx)$ with scalar product

$$\langle f, g\rangle = \int_a^b \overline{f(x)} g(x) w(x)\,dx.$$

In this space L is defined on sufficiently smooth functions which satisfy the above boundary conditions. Moreover, L gives rise to a self-adjoint operator. This can be seen formally by using integration by parts twice, where the boundary terms vanish by virtue of the boundary conditions. It then follows that the eigenvalues of a Sturm–Liouville operator are real and that eigenfunctions of L corresponding to different eigenvalues are orthogonal. However, this operator is unbounded and hence existence of an orthonormal basis of eigenfunctions is not evident. To overcome this problem, one looks at the resolvent

$$(L - z)^{-1}, \qquad z \in \mathbb{C},$$

where z is chosen to be some real number which is not an eigenvalue. Then, computing the resolvent amounts to solving the inhomogeneous equation, which can be done using the variation of parameters formula. This shows that the resolvent is an integral operator with a continuous symmetric kernel (the Green's function of the problem). As a consequence of the Arzelà–Ascoli theorem, this integral operator is compact and existence of a sequence of eigenvalues α_n which converge to 0 and eigenfunctions which form an orthonormal basis follows from the spectral theorem for compact operators. Finally, note that

$$(L - z)^{-1} u = \alpha u, \qquad Lu = \left(z + \alpha^{-1}\right) u,$$

are equivalent.

If the interval is unbounded, or if the coefficients have singularities at the boundary points, one calls L singular. In this case, the spectrum no longer consists of eigenvalues alone and can contain a continuous component. There is still an associated eigenfunction expansion (similar to Fourier series versus Fourier transform). This is important in quantum mechanics, since the one-dimensional time-independent Schrödinger equation is a special case of a S–L equation.

Example

We wish to find a function u(x) which solves the following Sturm–Liouville problem:

$$Lu = -\frac{d^2 u}{dx^2} = \lambda u$$

where the unknowns are λ and u(x). As above, we must add boundary conditions, we take for example

$$u(0) = u(\pi) = 0.$$

Observe that if *k* is any integer, then the function

$$u(x) = \sin kx$$

is a solution with eigenvalue $\lambda = k^2$. We know that the solutions of a S–L problem form an orthogonal basis, and we know from Fourier series that this set of sinusoidal functions is an orthogonal basis. Since orthogonal bases are always maximal (by definition) we conclude that the S–L problem in this case has no other eigenvectors.

Given the preceding, let us now solve the inhomogeneous problem

$$Lu = x, \qquad x \in (0, \pi)$$

with the same boundary conditions. In this case, we must write f(x) = x in a Fourier series. The reader may check, either by integrating $\int e^{ikx} x\, dx$ or by consulting a table of Fourier transforms, that we thus obtain

$$Lu = \sum_{k=1}^{\infty} -2\frac{(-1)^k}{k} \sin kx.$$

This particular Fourier series is troublesome because of its poor convergence properties. It is not clear a priori whether the series converges pointwise. Because of Fourier analysis, since the Fourier coefficients are "square-summable", the Fourier series converges in L^2 which is all we need for this particular theory to function. We mention for the interested reader that in this case we may rely on a result which says that Fourier's series converge at every point of differentiability, and at jump points (the function x, considered as a periodic function, has a jump at π) converges to the average of the left and right limits.

Therefore, by using formula ('4'), we obtain that the solution is

$$u = \sum_{k=1}^{\infty} 2\frac{(-1)^k}{k^3} \sin kx.$$

In this case, we could have found the answer using anti-differentiation. This technique yields

$$u = \frac{1}{6}\left(x^3 - \pi^2 x\right),$$

whose Fourier series agrees with the solution we found. The anti-differentiation technique is no longer useful in most cases when the differential equation is in many variables.

Application to Normal Modes

Certain partial differential equations can be solved with the help of S–L theory. Suppose we are interested in the modes of vibration of a thin membrane, held in a rectangular frame, $0 \leq x \leq L_1$, $0 \leq y \leq L_2$. The equation of motion for the vertical membrane's displacement, W(x, y, t) is given by the wave equation:

$$\frac{\partial^2 W}{\partial x^2} + \frac{\partial^2 W}{\partial y^2} = \frac{1}{c^2}\frac{\partial^2 W}{\partial t^2}.$$

The method of separation of variables suggests looking first for solutions of the simple form W = X(x) × Y(y) × T(t). For such a function W the partial differential equation becomes X"/X + Y"/Y = $(1/c^2)$T"/T. Since the three terms of this equation are functions of x,y,t separately, they must be constants. For example, the first term gives X" = λX for a constant λ. The boundary conditions ("held in a rectangular frame") are W = 0 when x = 0, L_1 or y = 0, L_2 and define the simplest possible S–L eigenvalue problems as in the example, yielding the "normal mode solutions" for W with harmonic time dependence,

$$W_{mn}(x,y,t) = A_{mn}\sin\left(\frac{m\pi x}{L_1}\right)\sin\left(\frac{n\pi y}{L_2}\right)\cos(\omega_{mn}t)$$

where m and n are non-zero integers, A_{mn} are arbitrary constants, and

$$\omega_{mn}^2 = c^2\left(\frac{m^2\pi^2}{L_1^2} + \frac{n^2\pi^2}{L_2^2}\right).$$

The functions W_{mn} form a basis for the Hilbert space of (generalized) solutions of the wave equation; that is, an arbitrary solution W can be decomposed into a sum of these modes, which vibrate at their individual frequencies ω_{mn}. This representation may require a convergent infinite sum.

Representation of Solutions and Numerical Calculation]

The Sturm–Liouville differential equation (1) with boundary conditions may be solved in practice by a variety of numerical methods. In difficult cases, one may need to carry out the intermediate calculations to several hundred decimal places of accuracy in order to obtain the eigenvalues correctly to a few decimal places.

1. Shooting methods. These methods proceed by guessing a value of λ, solving an initial value problem defined by the boundary conditions at one endpoint, say, a, of the interval [a, b], comparing the value this solution takes at the other endpoint b with the other desired boundary condition, and finally increasing or decreasing λ as necessary to correct the original value. This strategy is not applicable for locating complex eigenvalues.
2. Finite difference method.
3. The Spectral Parameter Power Series (SPPS) method makes use of a generalization of the following fact about second order ordinary differential equations: if y is a solution which does not vanish at any point of [a,b], then the function

$$y(x)\int_a^x \frac{dt}{p(t)y(t)^2}$$

is a solution of the same equation and is linearly independent from y. Further, all solutions are linear combinations of these two solutions. In the SPPS algorithm, one must begin with an arbitrary value λ_0^* (often $\lambda_0^* = 0$; it does not need to be an eigenvalue) and any solution y_0 of (1) with $\lambda = \lambda_0^*$ which does not vanish on [a, b]. (Discussion below of ways to find appropriate y_0 and λ_0^*.) Two sequences of functions $X^{(n)}(t)$, $X^{\sim(n)}(t)$ on [a, b], referred to as iterated integrals, are defined recursively as follows. First when n = 0, they are taken to be identically equal to 1 on [a, b]. To obtain the next functions they are multiplied alternately by $1/(py_0^2)$ and wy_0^2 and integrated, specifically

$$X^{(n)}(t) = \begin{cases} -\int_a^x X^{(n-1)}(t)p(t)^{-1}y_0(t)^{-2}\,dt & n \text{ odd,} \\ \int_a^x X^{(n-1)}(t)y_0(t)^2 w(t)\,dt & n \text{ even} \end{cases}$$

$$\widetilde{X}^{(n)}(t) = \begin{cases} \int_a^x \widetilde{X}^{(n-1)}(t)y_0(t)^2 w(t)\,dt & n \text{ odd,} \\ -\int_a^x \widetilde{X}^{(n-1)}(t)p(t)^{-1}y_0(t)^{-2}\,dt & n \text{ even} \end{cases}$$

when n > 0. The resulting iterated integrals are now applied as coefficients in the following two power series in λ:

$$u_0 = y_0 \sum_{k=0}^{\infty} \left(\lambda - \lambda_0^*\right)^k \widetilde{X}^{(2k)},$$

$$u_1 = y_0 \sum_{k=0}^{\infty} \left(\lambda - \lambda_0^*\right)^k X^{(2k+1)}.$$

Then for any λ (real or complex), u_0 and u_1 are linearly independent solutions of the corresponding equation (1). (The functions p(x) and q(x) take part in this construction through their influence on the choice of y_0.)

Next one chooses coefficients c_0, c_1 so that the combination $y = c_0u_0 + c_1u_1$ satisfies the first boundary condition (2). This is simple to do since $X^{(n)}(a) = 0$ and $X^{\sim(n)}(a) = 0$, for n > 0. The values of $X^{(n)}(b)$ and $X^{\sim(n)}(b)$ provide the values of $u_0(b)$ and $u_1(b)$ and the derivatives $u_0'(b)$ and $u_1'(b)$, so the second boundary condition (3) becomes an equation in a power series in λ. For numerical work one may truncate this series to a finite number of terms, producing a calculable polynomial in λ whose roots are approximations of the sought-after eigenvalues.

When $\lambda = \lambda_0$, this reduces to the original construction described above for a solution linearly independent to a given one. The representations ('5'),('6') also have theoretical applications in Sturm–Liouville theory.

Construction of a Nonvanishing Solution

The SPPS method can, itself, be used to find a starting solution y_0. Consider the equation $(py')' = \mu qy$; i.e., q, w, and λ are replaced in (1) by 0, −q, and μ respectively. Then the constant function 1 is a nonvanishing solution corresponding to the eigenvalue $\mu_0 = 0$. While there is no guarantee that u_0 or u_1 will not vanish, the complex function $y_0 = u_0 + iu_1$ will never vanish because two linearly independent solutions of a regular S–L equation cannot vanish simultaneously as a consequence of the Sturm separation theorem. This trick gives a solution y_0 of (1) for the value $\lambda_0 = 0$. In practice if (1) has real coefficients, the solutions based on y_0 will have very small imaginary parts which must be discarded.

Application to PDEs

For a linear second order in one spatial dimension and first order in time of the form:

$$f(x)\frac{\partial^2 u}{\partial x^2} + g(x)\frac{\partial u}{\partial x} + h(x)u = \frac{\partial u}{\partial t} + k(t)u$$

$$u(a,t) = u(b,t) = 0$$

$$u(x,0) = s(x)$$

Let us apply separation of variables, which in doing we must impose that:

$$u(x,t) = X(x)T(t)$$

Then our above PDE may be written as:

$$\frac{\hat{L}X(x)}{X(x)} = \frac{\hat{M}T(t)}{T(t)}$$

Where

$$\hat{L} = f(x)\frac{\mathrm{d}^2}{\mathrm{d}x^2} + g(x)\frac{\mathrm{d}}{\mathrm{d}x} + h(x), \qquad \hat{M} = \frac{\mathrm{d}}{\mathrm{d}t} + k(t)$$

Since, by definition, $\hat{L}$ and $X(x)$ are independent of time t and $\hat{M}$ and $T(t)$ are independent of position x, then both sides of the above equation must be equal to a constant:

$$\hat{L}X(x) = \lambda X(x)$$

$$X(a) = X(b) = 0$$

$$\hat{M}T(t) = \lambda T(t)$$

The first of these equations must be solved as a Sturm–Liouville problem. Since there is no general analytic (exact) solution to Sturm–Liouville problems, we can assume we already have the solution to this problem, that is, we have the eigenfunctions $X_n(x)$ and eigenvalues λ_n. The second of these equations can be analytically solved once the eigenvalues are known.

$$\frac{\mathrm{d}}{\mathrm{d}t}T_n(t) = (\lambda_n - k(t))T_n(t)$$

$$T_n(t) = a_n e^{-\left(\lambda_n t - \int_0^t k(\tau)\mathrm{d}\tau\right)}$$

$$u(x,t) = \sum_n a_n X_n(x) e^{-\left(\lambda_n t - \int_0^t k(\tau)\mathrm{d}\tau\right)}$$

$$a_n = \frac{\langle X_n(x), s(x)\rangle}{\langle X_n(x), X_n(x)\rangle}$$

Where:

$$\langle y(x), z(x)\rangle = \int_a^b y(x)z(x)w(x)\mathrm{d}x$$

$$w(x) = \frac{e^{\int \frac{g(x)}{f(x)}\mathrm{d}x}}{f(x)}$$

EXISTENCE AND UNIQUENESS OF SOLUTIONS

There are several theorems that establish existence and uniqueness of solutions to initial value problems involving ODEs both locally and globally. The two main theorems are

Theorem	Assumption	Conclusion
Peano existence theorem	F continuous	local existence only
Picard–Lindelöf theorem	F Lipschitz continuous	local existence and uniqueness

which are both local results.

Note that uniqueness theorems like the Lipschitz one above do not apply to DAE systems, which may have multiple solutions stemming from their (non-linear) algebraic part alone.

Local Existence and Uniqueness Theorem Simplified

The theorem can be stated simply as follows. For the equation and initial value problem:

$$y' = F(x, y), \quad y_0 = y(x_0)$$

if F and $\partial F/\partial y$ are continuous in a closed rectangle

$$R = [x_0 - a, x_0 + a] \times [y_0 - b, y_0 + b]$$

in the x-y plane, where a and b are real (symbolically: a, b $\in \mathbb{R}$) and $\times$ denotes the cartesian product, square brackets denote closed intervals, then there is an interval

$$I = [x_0 - h, x_0 + h] \subset [x_0 - a, x_0 + a]$$

for some h $\in \mathbb{R}$ where the solution to the above equation and initial value problem can be found. That is, there is a solution and it is unique. Since there is no restriction on F to be linear, this applies to non-linear equations that take the form F(x, y), and it can also be applied to systems of equations.

Global Uniqueness and Maximum Domain of Solution

When the hypotheses of the Picard–Lindelöf theorem are satisfied, then local existence and uniqueness can be extended to a global result. More precisely:

$$I_{\max} = (x_-, x_+), x_\pm \in \mathbb{R}, x_0 \in I_{\max}$$

For each initial condition (x_0, y_0) there exists a unique maximum (possibly infinite) open interval

such that any solution that satisfies this initial condition is a restriction of the solution that satisfies this initial condition with domain $I_{\max}$.

In the case that $x_\pm \nrightarrow \pm\infty$, there are exactly two possibilities

- explosion in finite time: $\limsup_{x \to x_\pm} \|y(x)\| \to \infty$
- leaves domain of definition: $\lim_{x \to x_\pm} y(x) \in \partial\bar{\Omega}$

where Ω is the open set in which F is defined, and $\partial\bar{\Omega}$ is its boundary.

Note that the maximum domain of the solution

- is always an interval (to have uniqueness)
- may be smaller than $\mathbb{R}$
- may depend on the specific choice of (x_0, y_0).

Example

$$y' = y^2$$

This means that F(x, y) = y^2, which is C^1 and therefore locally Lipschitz continuous, satisfying the Picard–Lindelöf theorem.

Even in such a simple setting, the maximum domain of solution cannot be all $\mathbb{R}$, since the solution is

$$y(x) = \frac{y_0}{(x_0 - x)y_0 + 1}$$

which has maximum domain:

$$\begin{cases} \mathbb{R} & y_0 = 0 \\ (-\infty, x_0 + \frac{1}{y_0}) & y_0 > 0 \\ (x_0 + \frac{1}{y_0}, +\infty) & y_0 < 0 \end{cases}$$

This shows clearly that the maximum interval may depend on the initial conditions. The domain of y could be taken as being $\mathbf{R} \setminus (x_0 + 1/y_0)$, but this would lead to a domain that is not an interval, so that the side opposite to the initial condition would be disconnected from the initial condition, and therefore not uniquely determined by it.

The maximum domain is not $\mathbb{R}$ because

$$\lim_{x \to x_\pm} \|y(x)\| \to \infty,$$

which is one of the two possible cases according to the above theorem.

REDUCTION OF ORDER

Differential equations can usually be solved more easily if the order of the equation can be reduced.

Reduction to a First-Order System

Any explicit differential equation of order n,

$$F\left(x, y, y', y'', \cdots, y^{(n-1)}\right) = y^{(n)}$$

can be written as a system of n first-order differential equations by defining a new family of unknown functions

$$y_i = y^{(i-1)}.$$

for i = 1, 2,... n. The n-dimensional system of first-order coupled differential equations is then

$$\begin{aligned} y_1' &= y_2 \\ y_2' &= y_3 \\ &\vdots \\ y_{n-1}' &= y_n \\ y_n' &= F(x, y_1, \cdots, y_n). \end{aligned}$$

more compactly in vector notation:

$$\mathbf{y}' = \mathbf{F}(x, \mathbf{y})$$

where

$$\mathbf{y} = (y_1, \cdots, y_n), \quad \mathbf{F}(x, y_1, \cdots, y_n) = (y_2, \cdots, y_n, F(x, y_1, \cdots, y_n)).$$

Summary of Exact Solutions

Some differential equations have solutions that can be written in an exact and closed form. Several important classes are given here.

In the table below, P(x), Q(x), P(y), Q(y), and M(x,y), N(x,y) are any integrable functions of x, y, and b and c are real given constants, and C_1, C_2,... are arbitrary constants (complex in general). The differential equations are in their equivalent and alternative forms that lead to the solution through integration.

In the integral solutions, λ and ε are dummy variables of integration (the continuum analogues of indices in summation), and the notation $\int^x F(\lambda)d\lambda$ just means to integrate F(λ) with respect to λ, then after the integration substitute λ = x, without adding constants (explicitly stated).

Differential equation	Solution method	General solution
	Separable equations	
First-order, separable in x and y (general case, see below for special cases) $P_1(x)Q_1(y) + P_2(x)Q_2(y)\frac{dy}{dx} = 0$ $P_1(x)Q_1(y)\,dx + P_2(x)Q_2(y)\,dy = 0$	Separation of variables (divide by P_2Q_1).	$\int^x \frac{P_1(\lambda)}{P_2(\lambda)}\,d\lambda + \int^y \frac{Q_2(\lambda)}{Q_1(\lambda)}\,d\lambda = C$
First-order, separable in x $\frac{dy}{dx} = F(x)$ $dy = F(x)\,dx$	Direct integration.	$y = \int^x F(\lambda)\,d\lambda + C$
First-order, autonomous, separable in y $\frac{dy}{dx} = F(y)$ $dy = F(y)\,dx$	Separation of variables (divide by F).	$x = \int^y \frac{d\lambda}{F(\lambda)} + C$
First-order, separable in x and y $P(y)\frac{dy}{dx} + Q(x) = 0$ $P(y)\,dy + Q(x)\,dx = 0$	Integrate throughout.	$\int^y P(\lambda)\,d\lambda + \int^x Q(\lambda)\,d\lambda = C$
	General first-order equations	
First-order, homogeneous $\frac{dy}{dx} = F\left(\frac{y}{x}\right)$	Set $y = ux$, then solve by separation of variables in u and x.	$\ln(Cx) = \int^{y/x} \frac{d\lambda}{F(\lambda) - \lambda}$
First-order, separable $yM(xy) + xN(xy)\frac{dy}{dx} = 0$ $yM(xy)\,dx + xN(xy)\,dy = 0$	Separation of variables (divide by xy).	$\ln(Cx) = \int^{xy} \frac{N(\lambda)\,d\lambda}{\lambda[N(\lambda) - M(\lambda)]}$ If $N = M$, the solution is $xy = C$.
Exact differential, first-order $M(x,y)\frac{dy}{dx} + N(x,y) = 0$ $M(x,y)\,dy + N(x,y)\,dx = 0$ where $\frac{\partial M}{\partial x} = \frac{\partial N}{\partial y}$	Integrate throughout.	$F(x,y) = \int^y M(x,\lambda)\,d\lambda + \int^x N(\lambda,y)\,d\lambda + Y(y) + X(x) = C$ where $Y(y)$ and $X(x)$ are functions from the integrals rather than constant values, which are set to make the final function $F(x, y)$ satisfy the initial equation.
Inexact differential, first-order $M(x,y)\frac{dy}{dx} + N(x,y) = 0$ $M(x,y)\,dy + N(x,y)\,dx = 0$ where $\frac{\partial M}{\partial x} \neq \frac{\partial N}{\partial y}$	Integration factor $\mu(x, y)$ satisfying $\frac{\partial(\mu M)}{\partial x} = \frac{\partial(\mu N)}{\partial y}$	If $\mu(x, y)$ can be found: $F(x,y) = \int^y \mu(x,\lambda)M(x,\lambda)\,d\lambda + \int^x \mu(\lambda,y)N(\lambda,y)\,d\lambda + Y(y) + X(x) = C$
	General second-order equations	

<table>
<tr>
<td>Second-order, autonomous
$\frac{d^2y}{dx^2} = F(y)$</td>
<td>Multiply equation by $2dy/dx$, substitute
$2\frac{dy}{dx}\frac{d^2y}{dx^2} = \frac{d}{dx}\left(\frac{dy}{dx}\right)^2$
, then integrate twice.</td>
<td>$x = \pm \int^y \frac{d\lambda}{\sqrt{2\int^\lambda F(\epsilon)\,d\epsilon + C_1}} + C_2$</td>
</tr>
<tr>
<td colspan="3">Linear equations (up to nth order)</td>
</tr>
<tr>
<td>First-order, linear, inhomogeneous, function coefficients
$\frac{dy}{dx} + P(x)y = Q(x)$</td>
<td>Integrating factor:
$e^{\int^x P(\lambda)\,d\lambda}$</td>
<td>$y = e^{-\int^x P(\lambda)\,d\lambda}\left[\int^x e^{\int^\lambda P(\epsilon)\,d\epsilon} Q(\lambda)\,d\lambda + C\right]$</td>
</tr>
<tr>
<td>Second-order, linear, inhomogeneous, constant coefficients
$\frac{d^2y}{dx^2} + b\frac{dy}{dx} + cy = r(x)$</td>
<td>Complementary function y_c: assume $y_c = e^{\alpha x}$, substitute and solve polynomial in α, to find the linearly independent functions $e^{\alpha_j x}$.

Particular integral y_p: in general the method of variation of parameters, though for very simple $r(x)$ inspection may work.</td>
<td>$y = y_c + y_p$
If $b^2 > 4c$, then:
$y_c = C_1 e^{\left(-b+\sqrt{b^2-4c}\right)\frac{x}{2}} + C_2 e^{-\left(b+\sqrt{b^2-4c}\right)\frac{x}{2}}$
If $b^2 = 4c$, then:
$y_c = (C_1 x + C_2)e^{-\frac{1}{2}bx}$
If $b^2 < 4c$, then:
$y_c = e^{-b\frac{x}{2}}\left[C_1 \sin\left(\sqrt{\lvert b^2-4c\rvert}\frac{x}{2}\right) + C_2 \cos\left(\sqrt{\lvert b^2-4c\rvert}\frac{x}{2}\right)\right]$</td>
</tr>
<tr>
<td>nth-order, linear, inhomogeneous, constant coefficients
$\sum_{j=0}^{n} b_j \frac{d^j y}{dx^j} = r(x)$</td>
<td>Complementary function y_c: assume $y_c = e^{\alpha x}$, substitute and solve polynomial in α, to find the linearly independent functions $e^{\alpha_j x}$.

Particular integral y_p: in general the method of variation of parameters, though for very simple $r(x)$ inspection may work.</td>
<td>$y = y_c + y_p$
Since α_j are the solutions of the polynomial of degree n:
$\prod_{j=1}^{n}(\alpha - \alpha_j) = 0$, then:
for α_j all different,
$y_c = \sum_{j=1}^{n} C_j e^{\alpha_j x}$
for each root α_j repeated k_j times,
$y_c = \sum_{j=1}^{n}\left(\sum_{\ell=1}^{k_j} C_\ell x^{\ell-1}\right) e^{\alpha_j x}$
for some α_j complex, then setting $\alpha = \chi_j + i\gamma_j$, and using Euler's formula, allows some terms in the previous results to be written in the form
$C_j e^{\alpha_j x} = C_j e^{\chi_j x}\cos(\gamma_j x + \phi_j)$
where ϕ_j is an arbitrary constant (phase shift).</td>
</tr>
</table>

REFERENCES

1. Halliday, David; Resnick, Robert (1977), Physics (3rd ed.), New York: Wiley, ISBN 0-471-71716-9
2. Harper, Charlie (1976), Introduction to Mathematical Physics, New Jersey: Prentice-Hall, ISBN 0-13-487538-9

3. Kreyszig, Erwin (1972), Advanced Engineering Mathematics (3rd ed.), New York: Wiley, ISBN 0-471-50728-8.
4. Polyanin, A. D. and V. F. Zaitsev, Handbook of Exact Solutions for Ordinary Differential Equations (2nd edition)", Chapman & Hall/CRC Press, Boca Raton, 2003. ISBN 1-58488-297-2
5. Simmons, George F. (1972), Differential Equations with Applications and Historical Notes, New York: McGraw-Hill, LCCN 75173716
6. Tipler, Paul A. (1991), Physics for Scientists and Engineers: Extended version (3rd ed.), New York: Worth Publishers, ISBN 0-87901-432-6
7. Boscain, Ugo; Chitour, Yacine (2011), Introduction à l'automatique (PDF) (in French)
8. Lawrence, Dresner (1999), Applications of Lie's Theory of Ordinary and Partial Differential Equations, Bristol and Philadelphia: Institute of Physics Publishing
9. Hazewinkel, Michiel, ed. (2001), "Sturm-Liouville theory", Encyclopedia of Mathematics, Springer, ISBN 978-1-55608-010-4
10. Hartman, Philip (2002). Ordinary Differential Equations (2 ed.). Philadelphia: SIAM. ISBN 978-0-89871-510-1.
11. Polyanin, A. D. & Zaitsev, V. F. (2003). Handbook of Exact Solutions for Ordinary Differential Equations (2 ed.). Boca Raton: Chapman & Hall/CRC Press. ISBN 1-58488-297-2.
12. Teschl, Gerald (2012). Ordinary Differential Equations and Dynamical Systems. Providence: American Mathematical Society. ISBN 978-0-8218-8328-0. (Chapter 5)
13. Teschl, Gerald (2009). Mathematical Methods in Quantum Mechanics; With Applications to Schrödinger Operators. Providence: American Mathematical Society. ISBN 978-0-8218-4660-5.
14. Zettl, Anton (2005). Sturm–Liouville Theory. Providence: American Mathematical Society. ISBN 0-8218-3905-5.
15. Birkhoff, Garrett (1973). A source book in classical analysis. Cambridge, Massachusetts: Harvard University Press. ISBN 0-674-82245-5.

CHAPTER 3

Partial Differential Equations

Chapter Contents

In mathematics, a partial differential equation (PDE) is a differential equation that contains unknown multivariable functions and their partial derivatives. (A special case are ordinary differential equations (ODEs), which deal with functions of a single variable and their derivatives.) PDEs are used to formulate problems involving functions of several variables, and are either solved by hand, or used to create a relevant computer model.

PDEs can be used to describe a wide variety of phenomena such as sound, heat, electrostatics, electrodynamics, fluid dynamics, elasticity, or quantum mechanics. These seemingly distinct physical phenomena can be formalised similarly in terms of PDEs. Just as ordinary differential equations often model one-dimensional dynamical systems, partial differential equations often model multidimensional systems. PDEs find their generalisation in stochastic partial differential equations.

INTRODUCTION

Partial differential equations (PDEs) are equations that involve rates of change with respect to continuous variables. The position of a rigid body is specified by six numbers, but the configuration of a fluid is given by the continuous distribution of several parameters, such as the temperature, pressure, and so forth. The dynamics for the rigid body take place in a finite-dimensional configuration space; the dynamics for the fluid occur in an infinite-dimensional configuration space. This distinction usually makes PDEs much harder to solve than ordinary differential equations (ODEs), but here again, there will be simple solutions for linear problems. Classic domains where PDEs are used include acoustics, fluid dynamics, electrodynamics, and heat transfer.

A partial differential equation (PDE) for the function $u(x_1, \cdots, x_n)$ is an equation of the form

$$f\left(x_1, \ldots, x_n, u, \frac{\partial u}{\partial x_1}, \ldots, \frac{\partial u}{\partial x_n}, \frac{\partial^2 u}{\partial x_1 \partial x_1}, \ldots, \frac{\partial^2 u}{\partial x_1 \partial x_n}, \ldots\right) = 0.$$

If f is a linear function of u and its derivatives, then the PDE is called linear. Common examples of linear PDEs include the heat equation, the wave equation, Laplace's equation, Helmholtz equation, Klein–Gordon equation, and Poisson's equation.

A relatively simple PDE is

$$u(x, y) = f(y),$$

This relation implies that the function $u(x,y)$ is independent of x. However, the equation gives no information on the function's dependence on the variable y. Hence the general solution of this equation is

$$\frac{du}{dx}(x) = 0,$$

where f is an arbitrary function of y. The analogous ordinary differential equation is

$$\frac{du}{dx}(x) = 0,$$

which has the solution

$$u(x) = c,$$

where c is any constant value. These two examples illustrate that general solutions of ordinary differential equations (ODEs) involve arbitrary constants, but solutions of PDEs involve arbitrary functions. A solution of a PDE is generally not unique; additional conditions must generally be specified on the boundary of the region where the solution is defined. For instance, in the simple example above, the function $f(y)$ can be determined if u is specified on the line $x = 0$.

EXISTENCE AND UNIQUENESS

Although the issue of existence and uniqueness of solutions of ordinary differential equations has a very satisfactory answer with the Picard–Lindelöf theorem, that is far from the case for partial differential equations. The Cauchy–Kowalevski theorem states that the Cauchy problem for any partial differential equation whose coefficients are analytic in the unknown function and its derivatives, has a locally unique analytic solution. Although this result might appear to settle the existence and uniqueness of solutions, there are examples of linear partial differential equations whose coefficients have derivatives of all orders (which are nevertheless not analytic) but which have no solutions at all: see Lewy (1957). Even if the solution of a partial differential equation exists and is unique, it may nevertheless have undesirable properties. The mathematical study of these questions is usually in the more powerful context of weak solutions.

An example of pathological behavior is the sequence (depending upon n) of Cauchy problems for the Laplace equation

$$\frac{\partial^2 u}{\partial x^2} + \frac{\partial^2 u}{\partial y^2} = 0,$$

with boundary conditions

$$u(x,0) = 0,$$
$$\frac{\partial u}{\partial y}(x,0) = \frac{\sin(nx)}{n},$$

where n is an integer. The derivative of u with respect to y approaches 0 uniformly in x as n increases, but the solution is

$$u(x,y) = \frac{\sinh(ny)\sin(nx)}{n^2}.$$

This solution approaches infinity if nx is not an integer multiple of π for any non-zero value of y. The Cauchy problem for the Laplace equation is called *ill-posed* or *not well-posed*, since the solution does not continuously depend on the data of the problem. Such ill-posed problems are not usually satisfactory for physical applications.

Notation

In PDEs, it is common to denote partial derivatives using subscripts. That is:

$$u_x = \frac{\partial u}{\partial x}$$
$$u_{xx} = \frac{\partial^2 u}{\partial x^2}$$
$$u_{xy} = \frac{\partial^2 u}{\partial y\,\partial x} = \frac{\partial}{\partial y}\left(\frac{\partial u}{\partial x}\right)$$

Especially in physics, del or Nabla (∇) is often used to denote spatial derivatives, and $\dot{u}\,\ddot{u}$ for time derivatives. For example, the wave equation (described below) can be written as

$$\ddot{u} = c^2\nabla^2 u$$

or

$$\ddot{u} = c^2\Delta u$$

where Δ is the Laplace operator.

CLASSIFICATION

Some linear, second-order partial differential equations can be classified as parabolic, hyperbolic and elliptic. Others such as the Euler–Tricomi equation have different types in different regions. The classification provides a guide to appropriate initial and boundary conditions, and to the smoothness of the solutions.

Equations of First Order

In mathematics, a first-order partial differential equation is a partial differential equation that involves only first derivatives of the unknown function of n variables. The equation takes the form

$$F(x_1, \ldots, x_n, u, u_{x_1}, \ldots u_{x_n}) = 0.$$

Such equations arise in the construction of characteristic surfaces for hyperbolic partial differential equations, in the calculus of variations, in some geometrical problems, and in simple models for gas dynamics whose solution involves the method of characteristics. If a family of solutions of a single first-order partial differential equation can be found, then additional solutions may be obtained by forming envelopes of solutions in that family. In a related procedure, general solutions may be obtained by integrating families of ordinary differential equations.

Characteristic Surfaces for the Wave Equation

Characteristic surfaces for the wave equation are level surfaces for solutions of the equation

$$u_t^2 = c^2 \left(u_x^2 + u_y^2 + u_z^2\right).$$

There is little loss of generality if we set $u_t = 1$: in that case u satisfies

$$u_x^2 + u_y^2 + u_z^2 = \frac{1}{c^2}.$$

In vector notation, let

$$\vec{x} = (x, y, z) \quad \text{and} \quad \vec{p} = (u_x, u_y, u_z).$$

A family of solutions with planes as level surfaces is given by

$$u(\vec{x}) = \vec{p} \cdot (\vec{x} - \vec{x_0}),$$

where

$$|\vec{p}\,| = \frac{1}{c}, \quad \text{and} \quad \vec{x_0} \quad \text{is arbitrary}.$$

If x and x_0 are held fixed, the envelope of these solutions is obtained by finding a point on the sphere of radius $1/c$ where the value of u is stationary. This is true if $\vec{p}$ is parallel to $\vec{x} - \vec{x_0}$. Hence the envelope has equation

$$u(\vec{x}) = \pm\frac{1}{c}|\vec{x} - \vec{x_0}\,|.$$

These solutions correspond to spheres whose radius grows or shrinks with velocity c. These are light cones in space-time.

The initial value problem for this equation consists in specifying a level surface S where u=0 for t=0. The solution is obtained by taking the envelope of all the spheres with centers on S, whose radii grow with velocity c. This envelope is obtained by requiring that

$$\frac{1}{c}|\vec{x} - \vec{x_0}| \quad \text{is stationary for} \quad \vec{x_0} \in S.$$

This condition will be satisfied if $|\vec{x} - \vec{x_0}|$ is normal to S. Thus the envelope corresponds to motion with velocity c along each normal to S. This is the Huygens' construction of wave fronts: each point on S emits a spherical wave at time t=0, and the wave front at a later time t is the envelope of these spherical waves. The normals to S are the light rays.

Two-Dimensional Theory

The notation is relatively simple in two space dimensions, but the main ideas generalize to higher dimensions. A general first-order partial differential equation has the form

$$F(x, y, u, p, q) = 0,$$

where

$$p = u_x, \quad q = u_y.$$

A complete integral of this equation is a solution $\varphi(x,y,u)$ that depends upon two parameters a and b. (There are n parameters required in the n-dimensional case.) An envelope of such solutions is obtained by choosing an arbitrary function w, setting b=$w(a)$, and determining $A(x,y,u)$ by requiring that the total derivative

$$\frac{d\varphi}{da} = \varphi_a(x, y, u, A, w(A)) + w'(A)\varphi_b(x, y, u, A, w(A)) = 0.$$

In that case, a solution u_w is also given by

$$u_w = \phi(x, y, u, A, w(A))$$

Each choice of the function w leads to a solution of the PDE. A similar process led to the construction of the light cone as a characteristic surface for the wave equation.

If a complete integral is not available, solutions may still be obtained by solving a system of ordinary equations. To obtain this system, first note that the PDE determines a cone (analogous to the light cone) at each point: if the PDE is linear in the derivatives of u (it is quasi-linear), then the cone degenerates into a line. In the general case, the pairs (p,q) that satisfy the equation determine a family of planes at a given point:

$$u - u_0 = p(x - x_0) + q(y - y_0),$$

where

$$F(x_0, y_0, u_0, p, q) = 0.$$

The envelope of these planes is a cone, or a line if the PDE is quasi-linear. The condition for an envelope is

$$F_p\,dp + F_q\,dq = 0,$$

where F is evaluated at (x_0, y_0, u_0, p, q), and dp and dq are increments of p and q that satisfy F=0. Hence the generator of the cone is a line with direction

$$dx : dy : du = F_p : F_q : (pF_p + qF_q).$$

This direction corresponds to the light rays for the wave equation. To integrate differential equations along these directions, we require increments for p and q along the ray. This can be obtained by differentiating the PDE:

$$F_x + F_u p + F_p p_x + F_q p_y = 0,$$
$$F_y + F_u q + F_p q_x + F_q q_y = 0,$$

Therefore the ray direction in (x, y, u, p, q) space is

$$dx : dy : du : dp : dq = F_p : F_q : (pF_p + qF_q) : (-F_x - F_u p) : (-F_y - F_u q).$$

The integration of these equations leads to a ray conoid at each point (x_0, y_0, u_0) General solutions of the PDE can then be obtained from envelopes of such conoids.

Linear Equations of Second Order

Assuming $u_{xy} = u_{yx}$, the general second-order PDE in two independent variables has the form

$$Au_{xx} + 2Bu_{xy} + Cu_{yy} + \cdots \text{(lower order terms)} = 0,$$

where the coefficients A, B, C etc. may depend upon x and y. If $A^2 + B^2 + C^2 > 0$ over a region of the xy plane, the PDE is second-order in that region. This form is analogous to the equation for a conic section:

$$Ax^2 + 2Bxy + Cy^2 + \cdots = 0.$$

More precisely, replacing ∂_x by X, and likewise for other variables (formally this is done by a Fourier transform), converts a constant-coefficient PDE into a polynomial of the same degree, with the top degree (a homogeneous polynomial, here a quadratic form) being most significant for the classification.

Just as one classifies conic sections and quadratic forms into parabolic, hyperbolic, and elliptic based on the discriminant $B^2 - 4AC$, the same can be done for a second-order PDE at a given point. However, the discriminant in a PDE is given by $B^2 - AC$ due to the convention of the xy term being $2B$ rather than B; formally, the discriminant (of the associated quadratic form) is $(2B)^2 - 4AC = 4(B^2 - AC)$ with the factor of 4 dropped for simplicity.

- $B^2 - AC < 0$: solutions of elliptic PDEs are as smooth as the coefficients allow, within the interior of the region where the equation and solutions are defined. For example, solutions of Laplace's equation are analytic within the domain where they are defined, but solutions may assume boundary values that are not smooth. The motion of a fluid at subsonic speeds can be approximated with elliptic PDEs, and the Euler–Tricomi equation is elliptic where $x < 0$.
- $B^2 - AC = 0$: equations that are parabolic at every point can be transformed into a form analogous to the heat equation by a change of independent variables. Solutions smooth out as the transformed time variable increases. The Euler–Tricomi equation has parabolic type on the line where $x = 0$.
- $B^2 - AC > 0$: hyperbolic equations retain any discontinuities of functions or derivatives in the initial data. An example is the wave equation. The motion of a fluid at supersonic speeds can be approximated with hyperbolic PDEs, and the Euler–Tricomi equation is hyperbolic where $x > 0$.

If there are n independent variables $x_1, x_2, \ldots, x_n$, a general linear partial differential equation of second order has the form

$$Lu = \sum_{i=1}^{n}\sum_{j=1}^{n} a_{i,j}\frac{\partial^2 u}{\partial x_i \partial x_j} \quad \text{plus lower-order terms} = 0.$$

The classification depends upon the signature of the eigenvalues of the coefficient matrix $a_{i,j}$..

- Elliptic: The eigenvalues are all positive or all negative.
- Parabolic : The eigenvalues are all positive or all negative, save one that is zero.
- Hyperbolic: There is only one negative eigenvalue and all the rest are positive, or there is only one positive eigenvalue and all the rest are negative.
- Ultrahyperbolic: There is more than one positive eigenvalue and more than one negative eigenvalue, and there are no zero eigenvalues. There is only a limited theory for ultra-hyperbolic equations (Courant and Hilbert, 1962).

Systems of First-Order Equations and Characteristic Surfaces

The classification of partial differential equations can be extended to systems of first-order equations, where the unknown u is now a vector with m components, and the coefficient matrices A_ν are m by m matrices for $\nu = 1, \ldots, n$. The partial differential equation takes the form

$$Lu = \sum_{\nu=1}^{n} A_\nu \frac{\partial u}{\partial x_\nu} + B = 0,$$

where the coefficient matrices A_ν and the vector B may depend upon x and u. If a hypersurface S is given in the implicit form

$$\varphi(x_1, x_2, \ldots, x_n) = 0,$$

where φ has a non-zero gradient, then S is a characteristic surface for the operator L at a given point if the characteristic form vanishes:

$$Q\left(\frac{\partial\varphi}{\partial x_1},\ldots,\frac{\partial\varphi}{\partial x_n}\right)=\det\left[\sum_{\nu=1}^{n}A_\nu\frac{\partial\varphi}{\partial x_\nu}\right]=0.$$

The geometric interpretation of this condition is as follows: if data for u are prescribed on the surface S, then it may be possible to determine the normal derivative of u on S from the differential equation. If the data on S and the differential equation determine the normal derivative of u on S, then S is non-characteristic. If the data on S and the differential equation *do not* determine the normal derivative of u on S, then the surface is characteristic, and the differential equation restricts the data on S: the differential equation is *internal* to S.

- A first-order system $Lu=0$ is *elliptic* if no surface is characteristic for L: the values of u on S and the differential equation always determine the normal derivative of u on S.
- A first-order system is *hyperbolic* at a point if there is a space-like surface S with normal ξ at that point. This means that, given any non-trivial vector η orthogonal to ξ, and a scalar multiplier λ, the equation $Q(\lambda\xi+\eta)=0$ has m real roots $\lambda_1, \lambda_2, \ldots, \lambda_m$. The system is strictly hyperbolic if these roots are always distinct. The geometrical interpretation of this condition is as follows: the characteristic form $Q(\zeta) = 0$ defines a cone (the normal cone) with homogeneous coordinates ζ. In the hyperbolic case, this cone has m sheets, and the axis $\zeta = \lambda\,\xi$ runs inside these sheets: it does not intersect any of them. But when displaced from the origin by η, this axis intersects every sheet. In the elliptic case, the normal cone has no real sheets.

Equations of Mixed Type

If a PDE has coefficients that are not constant, it is possible that it will not belong to any of these categories but rather be of **mixed type**. A simple but important example is the Euler–Tricomi equation

$u_{xx} = xu_{yy}$

which is called elliptic-hyperbolic because it is elliptic in the region $x < 0$, hyperbolic in the region $x > 0$, and degenerate parabolic on the line $x = 0$.

Infinite-Order PDEs in Quantum Mechanics

In the phase space formulation of quantum mechanics, one may consider the quantum Hamilton's equations for trajectories of quantum particles. These equations are infinite-order PDEs. However, in the semiclassical expansion, one has a finite system of ODEs at any fixed order of $\hbar$. The evolution equation of the Wigner function is also an infinite-order PDE. The quantum trajectories are quantum characteristics, with the use of which one could calculate the evolution of the Wigner function.

Analytical Solutions

Separation of Variables

Linear PDEs can be reduced to systems of ordinary differential equations by the important technique of separation of variables. This technique rests on a characteristic of solutions to differential equations: if one can find any solution that solves the equation and satisfies the boundary conditions, then it is *the* solution (this also applies to ODEs). We assume as an ansatz that the dependence of a solution on the parameters space and time can be written as a product of terms that each depend on a single parameter, and then see if this can be made to solve the problem.

In the method of separation of variables, one reduces a PDE to a PDE in fewer variables, which is an ordinary differential equation if in one variable – these are in turn easier to solve.

This is possible for simple PDEs, which are called separable partial differential equations, and the domain is generally a rectangle (a product of intervals). Separable PDEs correspond to diagonal matrices – thinking of "the value for fixed x" as a coordinate, each coordinate can be understood separately.

This generalizes to the method of characteristics, and is also used in integral transforms.

Method of Characteristics

In mathematics, the method of characteristics is a technique for solving partial differential equations. Typically, it applies to first-order equations, although more generally the method of characteristics is valid for any hyperbolic partial differential equation. The method is to reduce a partial differential equation to a family of ordinary differential equations along which the solution can be integrated from some initial data given on a suitable hypersurface.

Characteristics of First-Order Partial Differential Equation

For a first-order PDE (partial differential equation), the method of characteristics discovers curves (called characteristic curves or just characteristics) along which the PDE becomes an ordinary differential equation (ODE). Once the ODE is found, it can be solved along the characteristic curves and transformed into a solution for the original PDE.

For the sake of motivation, we confine our attention to the case of a function of two independent variables x and y for the moment. Consider a quasilinear PDE of the form

$$a(x,y,z)\frac{\partial z}{\partial x} + b(x,y,z)\frac{\partial z}{\partial y} = c(x,y,z).$$

Suppose that a solution z is known, and consider the surface graph $z = z(x,y)$ in $\mathbf{R}^3$. A normal vector to this surface is given by

$$\left(\frac{\partial z}{\partial x}(x,y), \frac{\partial z}{\partial y}(x,y), -1\right).$$

As a result, equation (**1**) is equivalent to the geometrical statement that the vector field

$$(a(x,y,z), b(x,y,z), c(x,y,z))$$

is tangent to the surface $z = z(x,y)$ at every point, for the dot product of this vector field with the above normal vector is zero. In other words, the graph of the solution must be a union of integral curves of this vector field. These integral curves are called the characteristic curves of the original partial differential equation.

The equations of the characteristic curve may be expressed invariantly by the *Lagrange-Charpit equations*

$$\frac{dx}{a(x,y,z)} = \frac{dy}{b(x,y,z)} = \frac{dz}{c(x,y,z)},$$

or, if a particular parametrization t of the curves is fixed, then these equations may be written as a system of ordinary differential equations for $x(t)$, $y(t)$, $z(t)$:

$$\begin{aligned} \frac{dx}{dt} &= a(x,y,z) \\ \frac{dy}{dt} &= b(x,y,z) \\ \frac{dz}{dt} &= c(x,y,z). \end{aligned}$$

These are the characteristic equations for the original system.

Linear and Quasilinear Cases

Consider now a PDE of the form

$$\sum_{i=1}^{n} a_i(x_1,\dots,x_n,u)\frac{\partial u}{\partial x_i} = c(x_1,\dots,x_n,u).$$

For this PDE to be linear, the coefficients a_i may be functions of the spatial variables only, and independent of u. For it to be quasilinear, a_i may also depend on the value of the function, but not on any derivatives. The distinction between these two cases is inessential for the discussion here.

For a linear or quasilinear PDE, the characteristic curves are given parametrically by

$$(x_1,\dots,x_n,u) = (x_1(s),\dots,x_n(s),u(s))$$

such that the following system of ODEs is satisfied

$$\frac{dx_i}{ds} = a_i(x_1, \ldots, x_n, u)$$

$$\frac{du}{ds} = c(x_1, \ldots, x_n, u).$$

These two equations give the characteristics of the PDE.

Fully Nonlinear Case

Consider the partial differential equation

$$F(x_1, \ldots, x_n, u, p_1, \ldots, p_n) = 0$$

where the variables p_i are shorthand for the partial derivatives

$$p_i = \frac{\partial u}{\partial x_i}.$$

Let $(x_i(s), u(s), p_i(s))$ be a curve in $\mathbf{R}^{2n+1}$. Suppose that u is any solution, and that

$$u(s) = u(x_1(s), \ldots, x_n(s)).$$

Along a solution, differentiating (4) with respect to s gives

$$\sum_i (F_{x_i} + F_u p_i)\dot{x}_i + \sum_i F_{p_i}\dot{p}_i = 0$$

$$\dot{u} - \sum_i p_i \dot{x}_i = 0$$

$$\sum_i (\dot{x}_i dp_i - \dot{p}_i dx_i) = 0.$$

The second equation follows from applying the chain rule to a solution u, and the third follows by taking an exterior derivative of the relation $du - \sum_i p_i \, dx_i = 0$. Manipulating these equations gives

$$\dot{x}_i = \lambda F_{p_i}, \quad \dot{p}_i = -\lambda(F_{x_i} + F_u p_i), \quad \dot{u} = \lambda \sum_i p_i F_{p_i}$$

where λ is a constant. Writing these equations more symmetrically, one obtains the Lagrange-Charpit equations for the characteristic

$$\frac{\dot{x}_i}{F_{p_i}} = -\frac{\dot{p}_i}{F_{x_i} + F_u p_i} = \frac{\dot{u}}{\sum p_i F_{p_i}}.$$

Geometrically, the method of characteristics in the fully nonlinear case can be interpreted as requiring that the Monge cone of the differential equation should everywhere be tangent to the graph of the solution.

Example

As an example, consider the advection equation (this example assumes familiarity with PDE notation, and solutions to basic ODEs).

$$a\frac{\partial u}{\partial x}+\frac{\partial u}{\partial t}=0$$

where a is constant and u is a function of x and t. We want to transform this linear first-order PDE into an ODE along the appropriate curve; i.e. something of the form

$$\frac{d}{ds}u(x(s),t(s))=F(u,x(s),t(s)),$$

where $(x(s),t(s))$ is a characteristic line. First, we find

$$\frac{d}{ds}u(x(s),t(s))=\frac{\partial u}{\partial x}\frac{dx}{ds}+\frac{\partial u}{\partial t}\frac{dt}{ds}$$

by the chain rule. Now, if we set $\frac{dx}{ds}=a$ and $\frac{dt}{ds}=1$ we get

$$a\frac{\partial u}{\partial x}+\frac{\partial u}{\partial t}$$

which is the left hand side of the PDE we started with. Thus

$$\frac{d}{ds}u=a\frac{\partial u}{\partial x}+\frac{\partial u}{\partial t}=0.$$

So, along the characteristic line $(x(s),t(s))$, the original PDE becomes the ODE $u_s=F(u,x(s),t(s))=0$. That is to say that along the characteristics, the solution is constant. Thus, $u(x_s,t_s)=u(x_0,0)$ where (x_s,t_s) and $(x_0,0)$ lie on the same characteristic. So to determine the general solution, it is enough to find the characteristics by solving the characteristic system of ODEs:

$\frac{dt}{ds}=1$, letting $t(0)=0$ we know $t=s$,

$\frac{dx}{ds}=a$, letting $x(0)=x_0$ we know $x=as+x_0=at+x_0$,

$\frac{du}{ds}=0$, letting $u(0)=f(x_0)$ we know

$u(x(t),t)=f(x_0)=f(x-at)$.

In this case, the characteristic lines are straight lines with slope a, and the value of u remains constant along any characteristic line.

Characteristics of Linear Differential Operators

Let X be a differentiable manifold and P a linear differential operator

$$P : C^\infty(X) \to C^\infty(X)$$

of order k. In a local coordinate system x^i,

$$P = \sum_{|\alpha| \leq k} P^\alpha(x) \frac{\partial}{\partial x^\alpha}$$

in which α denotes a multi-index. The principal symbol of P, denoted σ_P, is the function on the cotangent bundle T^*X defined in these local coordinates by

$$\sigma_P(x, \xi) = \sum_{|\alpha| = k} P^\alpha(x) \xi_\alpha$$

where the ξ_i are the fiber coordinates on the cotangent bundle induced by the coordinate differentials dx^i. Although this is defined using a particular coordinate system, the transformation law relating the ξ_i and the x^i ensures that σ_P is a well-defined function on the cotangent bundle.

The function σ_P is homogeneous of degree k in the ξ variable. The zeros of σ_P, away from the zero section of T^*X, are the characteristics of P. A hypersurface of X defined by the equation $F(x) = c$ is called a characteristic hypersurface at x if

$$\sigma_P(x, dF(x)) = 0.$$

Invariantly, a characteristic hypersurface is a hypersurface whose conormal bundle is in the characteristic set of P.

Qualitative Analysis of Characteristics

Characteristics are also a powerful tool for gaining qualitative insight into a PDE.

One can use the crossings of the characteristics to find shock waves for potential flow in a compressible fluid. Intuitively, we can think of each characteristic line implying a solution to u along itself. Thus, when two characteristics cross, the function becomes multi-valued resulting in a non-physical solution. Physically, this contradiction is removed by the formation of a shock wave, a tangential discontinuity or a weak discontinuity and can result in non-potential flow, violating the initial assumptions.

Characteristics may fail to cover part of the domain of the PDE. This is called a rarefaction, and indicates the solution typically exists only in a weak, i.e. integral equation, sense.

The direction of the characteristic lines indicate the flow of values through the solution, as the example above demonstrates. This kind of knowledge is useful when solving PDEs numerically as it can indicate which finite difference scheme is best for the problem.

Integral Transform

An integral transform may transform the PDE to a simpler one, in particular, a separable PDE. This corresponds to diagonalizing an operator.

An important example of this is Fourier analysis, which diagonalizes the heat equation using the eigenbasis of sinusoidal waves.

If the domain is finite or periodic, an infinite sum of solutions such as a Fourier series is appropriate, but an integral of solutions such as a Fourier integral is generally required for infinite domains. The solution for a point source for the heat equation given above is an example of the use of a Fourier integral.

Change of Variables

Often a PDE can be reduced to a simpler form with a known solution by a suitable change of variables. For example, the Black–Scholes PDE

$$\frac{\partial V}{\partial t}+\frac{1}{2}\sigma^2 S^2\frac{\partial^2 V}{\partial S^2}+rS\frac{\partial V}{\partial S}-rV=0$$

is reducible to the heat equation

$$\frac{\partial u}{\partial \tau}=\frac{\partial^2 u}{\partial x^2}$$

by the change of variables (for complete details see Solution of the Black Scholes Equation at the Wayback Machine (archived April 11, 2008))

$$V(S,t)=Kv(x,\tau)$$
$$x=\ln\left(\tfrac{S}{K}\right)$$
$$\tau=\tfrac{1}{2}\sigma^2(T-t)$$
$$v(x,\tau)=\exp(-\alpha x-\beta\tau)u(x,\tau).$$

Fundamental Solution

In mathematics, a fundamental solution for a linear partial differential operator L is a formulation in the language of distribution theory of the older idea of a Green's function (which normally further addresses boundary conditions).

In terms of the Dirac delta "function" $\delta(x)$, a fundamental solution F is the solution of the inhomogeneous equation

$$LF=\delta(x)\ .$$

Here F is *a priori* only assumed to be a distribution.

This concept has long been utilized for the Laplacian in two and three dimensions. (It was investigated for all dimensions for the Laplacian by Marcel Riesz.)

The existence of a fundamental solution for any operator with constant coefficients — the most important case, directly linked to the possibility of using convolution to solve an arbitrary right hand side — was shown by Bernard Malgrange and Leon Ehrenpreis.

Example

Consider the following differential equation $Lf = \sin(x)$ with

$$L = \frac{d^2}{dx^2}.$$

The fundamental solutions can be obtained by solving $LF = \delta(x)$, explicitly,

$$\frac{d^2}{dx^2}F(x) = \delta(x) .$$

Since for the Heaviside function H we have

$$\frac{d}{dx}H(x) = \delta(x) ,$$

there is a solution

$$\frac{d}{dx}F(x) = H(x) + C .$$

Here C is an arbitrary constant introduced by the integration. For convenience, set $C = -1/2$.

After integrating ${}^{dF}/_{dx}$ and choosing the new integration constant as zero, one has

$$F(x) = xH(x) - \frac{1}{2}x = \frac{1}{2}|x| .$$

Motivation

Once the fundamental solution is found, it is easy to find the desired solution of the original equation. In fact, this process is achieved by convolution.

Fundamental solutions also play an important role in the numerical solution of partial differential equations by the boundary element method.

Application to the Example

Consider the operator L and the differential equation mentioned in the example,

$$\frac{d^2}{dx^2} f(x) = \sin(x) .$$

We can find the solution of the original equation by convolving the right-hand side $\sin(x)$ with the fundamental solution $F(x) = |x|/2$,

$$f(x) = \int_{-\infty}^{\infty} \frac{1}{2} |x - y| \sin(y) dy .$$

This shows that some care must be taken when working with functions which do not have enough regularity (e.g. compact support, L^1 integrability) since, we know that the desired solution is $f(x) = -\sin x$, while the above integral diverges for all x. The two expressions for f are, however, equal as distributions.

An Example That More Clearly Works

$$\frac{d^2}{dx^2} f(x) = I(x) ,$$

where I is the characteristic (indicator) function of the unit interval $[0,1]$. In that case, it can be readily verified that the convolution $I*F$ with $F(x)=|x|/2$ is a solution, i.e., has second derivative equal to I.

Proof that the Convolution is a Solution

Denote the convolution of functions F and g as $F*g$. Say we are trying to find the solution of $Lf = g(x)$. We want to prove that $F*g$ is a solution of the previous equation, i.e. we want to prove that $L(F*g) = g$. When applying the differential operator, L, to the convolution, it is known that

$$L(F * g) = (LF) * g ,$$

provided L has constant coefficients.

If F is the fundamental solution, the right side of the equation reduces to

$$\delta * g$$

But since the delta function is an identity element for convolution, this is simply $g(x)$. Summing up,

$$L(F * g) = (LF) * g = \delta(x) * g(x) = \int_{-\infty}^{\infty} \delta(x - y) g(y) dy = g(x) .$$

Therefore, if F is the fundamental solution, the convolution $F*g$ is one solution of $Lf = g(x)$. This does not mean that it is the only solution. Several solutions for different initial conditions can be found.

Fundamental Solutions for Some Partial Differential Equations

Laplace Equation

For the Laplace equation,

$$[-\Delta]\Phi(\mathbf{x},\mathbf{x}') = \delta(\mathbf{x}-\mathbf{x}')$$

the fundamental solutions in two and three dimensions, respectively, are

$$\Phi_{2D}(\mathbf{x},\mathbf{x}') = -\frac{1}{2\pi}\ln|\mathbf{x}-\mathbf{x}'|, \qquad \Phi_{3D}(\mathbf{x},\mathbf{x}') = \frac{1}{4\pi|\mathbf{x}-\mathbf{x}'|}.$$

Screened Poisson Equation

For the screened Poisson equation,

$$[-\Delta + k^2]\Phi(\mathbf{x},\mathbf{x}') = \delta(\mathbf{x}-\mathbf{x}'), \quad k \in \mathbb{R},$$

the fundamental solutions are

$$\Phi_{2D}(\mathbf{x},\mathbf{x}') = \frac{1}{2\pi}K_0(k|\mathbf{x}-\mathbf{x}'|), \qquad \Phi_{3D}(\mathbf{x},\mathbf{x}') = \frac{\exp(-k|\mathbf{x}-\mathbf{x}'|)}{4\pi|\mathbf{x}-\mathbf{x}'|},$$

where K_0 is a modified Bessel function of the second kind.

Biharmonic Equation

For the Biharmonic equation,

$$[-\Delta^2]\Phi(\mathbf{x},\mathbf{x}') = \delta(\mathbf{x}-\mathbf{x}')$$

the biharmonic equation has the fundamental solutions

$$\Phi_{2D}(\mathbf{x},\mathbf{x}') = -\frac{|\mathbf{x}-\mathbf{x}'|^2}{8\pi}(\ln|\mathbf{x}-\mathbf{x}'|-1), \qquad \Phi_{3D}(\mathbf{x},\mathbf{x}') = \frac{|\mathbf{x}-\mathbf{x}'|}{8\pi}.$$

Signal Processing

In signal processing, the analog of the fundamental solution of a differential equation is called the impulse response of a filter.

Superposition Principle

Because any superposition of solutions of a linear, homogeneous PDE is again a solution, the particular solutions may then be combined to obtain more general solutions. if u1 and u2 are solutions of a homogeneous linear pde in same region R, then u= c1u1+c2u2 with any constants c1 and c2 are also a solution of that pde in that same region....

Methods for Non-Linear Equations

There are no generally applicable methods to solve nonlinear PDEs. Still, existence and uniqueness results (such as the Cauchy–Kowalevski theorem) are often possible, as are proofs of important qualitative and quantitative properties of solutions (getting these results is a major part of analysis). Computational solution to the nonlinear PDEs, the split-step method, exist for specific equations like nonlinear Schrödinger equation.

Nevertheless, some techniques can be used for several types of equations. The h-principle is the most powerful method to solve underdetermined equations. The Riquier–Janet theory is an effective method for obtaining information about many analytic overdetermined systems.

The method of characteristics (similarity transformation method) can be used in some very special cases to solve partial differential equations.

In some cases, a PDE can be solved via perturbation analysis in which the solution is considered to be a correction to an equation with a known solution. Alternatives are numerical analysis techniques from simple finite difference schemes to the more mature multigrid and finite element methods. Many interesting problems in science and engineering are solved in this way using computers, sometimes high performance supercomputers.

Lie Group Method

From 1870 Sophus Lie's work put the theory of differential equations on a more satisfactory foundation. He showed that the integration theories of the older mathematicians can, by the introduction of what are now called Lie groups, be referred to a common source; and that ordinary differential equations which admit the same infinitesimal transformations present comparable difficulties of integration. He also emphasized the subject of transformations of contact.

A general approach to solving PDE's uses the symmetry property of differential equations, the continuous infinitesimal transformations of solutions to solutions (Lie theory). Continuous group theory, Lie algebras and differential geometry are used to understand the structure of linear and nonlinear partial differential equations for generating integrable equations, to find its Lax pairs, recursion operators, Bäcklund transform and finally finding exact analytic solutions to the PDE.

Symmetry methods have been recognized to study differential equations arising in mathematics, physics, engineering, and many other disciplines.

Semianalytical Methods

The adomian decomposition method, the Lyapunov artificial small parameter method, and He's homotopy perturbation method are all special cases of the more general homotopy analysis method. These are series expansion methods, and except for the Lyapunov method, are independent of small physical parameters as compared to the well known perturbation theory, thus giving these methods greater flexibility and solution generality.

Numerical Solutions

The three most widely used numerical methods to solve PDEs are the finite element method (FEM), finite volume methods (FVM) and finite difference methods (FDM). The FEM has a prominent position among these methods and especially its exceptionally efficient higher-order version hp-FEM. Other versions of FEM include the generalized finite element method (GFEM), extended finite element method (XFEM), spectral finite element method (SFEM), meshfree finite element method, discontinuous Galerkin finite element method (DGFEM), Element-Free Galerkin Method (EFGM), Interpolating Element-Free Galerkin Method (IEFGM), etc.

Finite Element Method

The finite element method (FEM) is a numerical technique for finding approximate solutions to boundary value problems for partial differential equations. It is also referred to as finite element analysis (FEA). It subdivides a large problem into smaller, simpler parts that are called finite elements. The simple equations that model these finite elements are then assembled into a larger system of equations that models the entire problem. FEM then uses variational methods from the calculus of variations to approximate a solution by minimizing an associated error function.

Basic Concepts

The subdivision of a whole domain into simpler parts has several advantages:

- Accurate representation of complex geometry
- Inclusion of dissimilar material properties
- Easy representation of the total solution
- Capture of local effects.

A typical work out of the method involves (1) dividing the domain of the problem into a collection of subdomains, with each subdomain represented by a set of element equations to the original problem, followed by (2) systematically recombining all sets of element equations into a global system of equations for the final calculation. The global system of equations has known solution techniques, and can be calculated from the initial values of the original problem to obtain a numerical answer.

In the first step above, the element equations are simple equations that locally approximate the original complex equations to be studied, where the original equations are often partial differential equations (PDE). To explain the approximation in this process, FEM is commonly introduced as a special case of Galerkin method. The process, in mathematical language, is to construct an integral of the inner product of the residual and the weight functions and set the integral to zero. In simple terms, it is a procedure that minimizes the error of approximation by fitting trial functions into the PDE. The residual is the error caused by the trial functions, and the weight functions are polynomial approximation functions that project the residual. The process eliminates all the spatial derivatives from the PDE, thus approximating the PDE locally with

- a set of algebraic equations for steady state problems,
- a set of ordinary differential equations for transient problems.

These equation sets are the element equations. They are linear if the underlying PDE is linear, and vice versa. Algebraic equation sets that arise in the steady state problems are solved using numerical linear algebra methods, while ordinary differential equation sets that arise in the transient problems are solved by numerical integration using standard techniques such as Euler's method or the Runge-Kutta method.

In step (2) above, a global system of equations is generated from the element equations through a transformation of coordinates from the subdomains' local nodes to the domain's global nodes. This spatial transformation includes appropriate orientation adjustments as applied in relation to the reference coordinate system. The process is often carried out by FEM software using coordinate data generated from the subdomains.

FEM is best understood from its practical application, known as finite element analysis (FEA). FEA as applied in engineering is a computational tool for performing engineering analysis. It includes the use of mesh generation techniques for dividing a complex problem into small elements, as well as the use of software program coded with FEM algorithm. In applying FEA, the complex problem is usually a physical system with the underlying physics such as the Euler-Bernoulli beam equation, the heat equation, or the Navier-Stokes equations expressed in either PDE or integral equations, while the divided small elements of the complex problem represent different areas in the physical system.

FEM mesh created by an analyst prior to finding a solution to a magnetic problem using FEM software. Colours indicate that the analyst has set material properties for each zone, in this case a conducting wire coil in orange; a ferromagnetic component (perhaps iron) in light blue; and air in grey. Although the geometry may seem simple, it would be very challenging to calculate the magnetic field for this setup without FEM software, using equations alone.

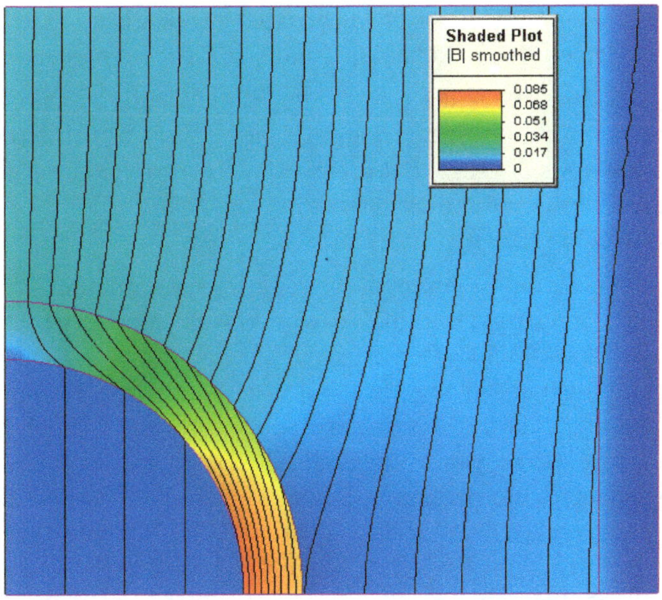

FEM solution to the problem at left, involving a cylindrically shaped magnetic shield. The ferromagnetic cylindrical part is shielding the area inside the cylinder by diverting the magnetic field created by the coil (rectangular area on the right). The color represents the amplitude of the magnetic flux density, as indicated by the scale in the inset legend, red being high amplitude. The area inside the cylinder is low amplitude (dark blue, with widely spaced lines of magnetic flux), which suggests that the shield is performing as it was designed to.

FEA is a good choice for analyzing problems over complicated domains (like cars and oil pipelines), when the domain changes (as during a solid state reaction with a moving boundary), when the desired precision varies over the entire domain, or when the solution lacks smoothness. For instance, in a frontal crash simulation it is possible to increase prediction accuracy in "important" areas like the front of the car and reduce it in its rear (thus reducing cost of the simulation). Another example would be in numerical weather prediction, where it is more important to have accurate predictions over developing highly nonlinear phenomena (such as tropical cyclones in the atmosphere, or eddies in the ocean) rather than relatively calm areas.

History

While it is difficult to quote a date of the invention of the finite element method, the method originated from the need to solve complex elasticity and structural analysis problems in civil and aeronautical engineering. Its development can be traced back to the work by A. Hrennikoff and R. Courant in the early 1940s. In China, in the later 1950s and early 1960s, based on the computations of dam constructions, K. Feng proposed a systematic numerical method for solving partial differential equations. The method was called the finite difference method based on variation principle, which was another independent invention of the finite element method. Although the approaches used by these pioneers are different, they share one essential characteristic: mesh discretization of a continuous domain into a set of discrete sub-domains, usually called elements.

Hrennikoff's work discretizes the domain by using a lattice analogy, while Courant's approach divides the domain into finite triangular subregions to solve second order elliptic partial differential equations (PDEs) that arise from the problem of torsion of a cylinder. Courant's contribution was evolutionary, drawing on a large body of earlier results for PDEs developed by Rayleigh, Ritz, and Galerkin.

The finite element method obtained its real impetus in the 1960s and 1970s by the developments of J. H. Argyris with co-workers at the University of Stuttgart, R. W. Clough with co-workers at UC Berkeley, O. C. Zienkiewicz with co-workers Ernest Hinton, Bruce Irons and others at the University of Swansea, Philippe G. Ciarlet at the University of Paris 6 and Richard Gallagher with co-workers at Cornell University. Further impetus was provided in these years by available open source finite element software programs. NASA sponsored the original version of NASTRAN, and UC Berkeley made the finite element program SAP IV widely available. In Norway the ship classification society Det Norske Veritas (now DNV GL) developed Sesam in 1969 for use in analysis of ships. A rigorous mathematical basis to the finite element method was provided in 1973 with the publication by Strang and Fix. The method has since been generalized for the numerical modeling of physical systems in a wide variety of engineering disciplines, e.g., electromagnetism, heat transfer, and fluid dynamics.

Technical Discussion

The Structure of Finite Element Methods

Finite element methods are numerical methods for approximating the solutions of mathematical problems that are usually formulated so as to precisely state an idea of some aspect of physical reality.

A finite element method is characterized by a variational formulation, a discretization strategy, one or more solution algorithms and post-processing procedures.

Examples of variational formulation are the Galerkin method, the discontinuous Galerkin method, mixed methods, etc.

A discretization strategy is understood to mean a clearly defined set of procedures that cover (a) the creation of finite element meshes, (b) the definition of basis function on reference elements (also called shape functions) and (c) the mapping of reference elements onto the elements of the mesh. Examples of discretization strategies are the h-version, p-version, hp-version, x-FEM, isogeometric analysis, etc. Each discretization strategy has certain advantages and disadvantages. A reasonable criterion in selecting a discretization strategy is to realize nearly optimal performance for the broadest set of mathematical models in a particular model class.

There are various numerical solution algorithms that can be classified into two broad categories; direct and iterative solvers. These algorithms are designed to exploit the sparsity of matrices that depend on the choices of variational formulation and discretization strategy.

Postprocessing procedures are designed for the extraction of the data of interest from a finite element solution. In order to meet the requirements of solution verification, postprocessors need to provide for *a posteriori* error estimation in terms of the quantities of interest. When the errors of approximation are larger than what is considered acceptable then the discretization has to be changed either by an automated adaptive process or by action of the analyst. There are some very efficient postprocessors that provide for the realization of superconvergence.

Illustrative Problems P1 and P2

We will illustrate the finite element method using two sample problems from which the general method can be extrapolated. It is assumed that the reader is familiar with calculus and linear algebra.

P1 is a one-dimensional problem

$$\text{P1}\ :\begin{cases} u''(x) = f(x) \text{ in } (0,1), \\ u(0) = u(1) = 0, \end{cases}$$

where f is given, u is an unknown function of x, and u'' is the second derivative of u with respect to x.

P2 is a two-dimensional problem (Dirichlet problem)

$$\text{P2}\ :\begin{cases} u_{xx}(x,y) + u_{yy}(x,y) = f(x,y) & \text{in } \Omega, \\ u = 0 & \text{on } \partial\Omega, \end{cases}$$

where Ω is a connected open region in the xy plane whose boundary $\partial\Omega$ is "nice" (e.g., a smooth manifold or a polygon), and u_{xx} and u_{yy} denote the second derivatives with respect to x and y, respectively.

The problem P1 can be solved "directly" by computing antiderivatives. However, this method of solving the boundary value problem (BVP) works only when there is one spatial dimension and does not generalize to higher-dimensional problems or to problems like $u + u'' = f$. For this reason, we will develop the finite element method for P1 and outline its generalization to P2.

Our explanation will proceed in two steps, which mirror two essential steps one must take to solve a boundary value problem (BVP) using the FEM.

- In the first step, one rephrases the original BVP in its weak form. Little to no computation is usually required for this step. The transformation is done by hand on paper.
- The second step is the discretization, where the weak form is discretized in a finite-dimensional space.

After this second step, we have concrete formulae for a large but finite-dimensional linear problem whose solution will approximately solve the original BVP. This finite-dimensional problem is then implemented on a computer.

Weak Formulation

The first step is to convert P1 and P2 into their equivalent weak formulations.

The Weak form of P1

If u solves P1, then for any smooth function v that satisfies the displacement boundary conditions, i.e. $v = 0$ at $x = 0$ and $x = 1$, we have

(1) $$\int_0^1 f(x)v(x)\,dx = \int_0^1 u''(x)v(x)\,dx.$$

Conversely, if u with $u(0) = u(1) = 0$ satisfies (1) for every smooth function v(x) then one may show that this u will solve P1. The proof is easier for twice continuously differentiable u (mean value theorem), but may be proved in a distributional sense as well.

We define a new function $\phi(u,v)$ by using integration by parts on the right-hand-side of (1):

(2) $$\begin{aligned}\int_0^1 f(x)v(x)\,dx &= \int_0^1 u''(x)v(x)\,dx \\ &= u'(x)v(x)|_0^1 - \int_0^1 u'(x)v'(x)\,dx \\ &= -\int_0^1 u'(x)v'(x)\,dx \equiv -\phi(u,v),\end{aligned}$$

where we have used the assumption that $v(0) = v(1) = 0$.

The Weak form of P2

If we integrate by parts using a form of Green's identities, we see that if u solves P2, then we may define $\phi(u,v)$ for any v by

$$\int_\Omega fv\,ds = -\int_\Omega \nabla u \cdot \nabla v\,ds \equiv -\phi(u,v),$$

where ∇ denotes the gradient and $\cdot$ denotes the dot product in the two-dimensional plane. Once more ϕ can be turned into an inner product on a suitable space $H^0_1(\Omega)$ of "once differentiable" functions of Ω that are zero on $\partial\Omega$. We have also assumed that $v \in H_0^1(\Omega)$ (see Sobolev spaces). Existence and uniqueness of the solution can also be shown.

A Proof Outline of Existence And Uniqueness of the Solution

We can loosely think of $H_0^1(0,1)$ to be the absolutely continuous functions of (0,1) that are 0 at x = 0 and x = 1 (see Sobolev spaces). Such functions are (weakly) "once differentiable" and it turns out that the symmetric bilinear map ϕ then defines an inner product which turns $H_0^1(0,1)$ into a Hilbert space (a detailed proof is nontrivial). On the other hand, the left-hand-side $\int_0^1 f(x)v(x)dx$ is also an inner product, this time on the Lp space $L^2(0,1)$. An application of the Riesz representation theorem for Hilbert spaces shows that there is a unique u solving (2) and therefore P1. This solution is a-priori only a member of $H_0^1(0,1)$ but using elliptic regularity, will be smooth if f is.

Discretization

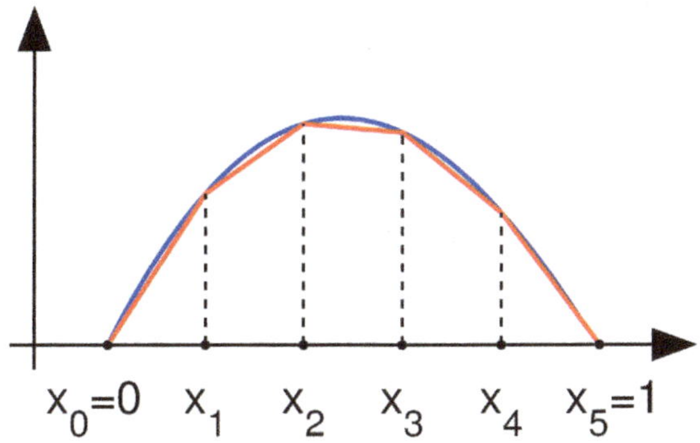

A function in {\displaystyle H_{0}^{1},} with zer o values at the endpoints (blue), and a piecewise linear approximation (red)

P1 and P2 are ready to be discretized which leads to a common sub-problem (3). The basic idea is to replace the infinite-dimensional linear problem:

Find $u \in H_0^1$ such that

$$\forall v \in H_0^1,\ -\phi(u,v) = \int fv$$

with a finite-dimensional version:

(3) Find $u \in V$ such that

$$\forall v \in V,\ -\phi(u,v) = \int fv$$

where V is a finite-dimensional subspace of H_0^1. There are many possible choices for V (one possibility leads to the spectral method). However, for the finite element method we take V to be a space of piecewise polynomial functions.

For Problem P1

We take the interval (0,1), choose n values of x with 0 = x 0 < x 1 < ⋯ < x n < x n + 1 = 1 and we define V by: $V = \{v : [0,1] \to \mathbb{R}\ : v \text{ is continuous, } v|_{[x_k,x_{k+1}]}$ $\text{is linear for } k = 0,\dots,n, \text{ and } v(0) = v(1) = 0\}$ where we define $x_0 = 0$ and $x_{n+1} = 1$. Observe that functions in V are not differentiable according to the elementary definition of calculus. Indeed, if $v \in V$ then the derivative is typically not defined at any $x = x_k$, $k = 1, \ldots, n$. However, the derivative exists at every other value of x and one can use this derivative for the purpose of integration by parts.

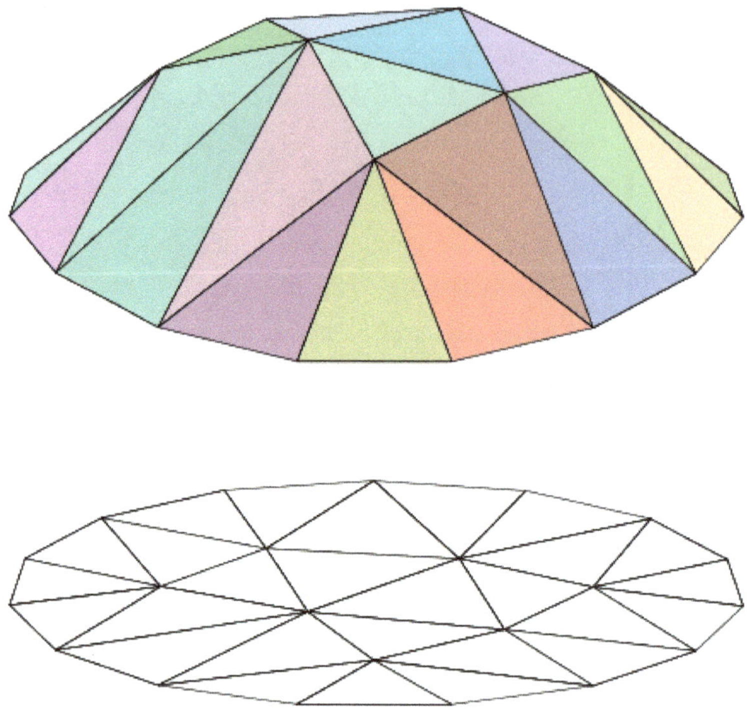

A piecewise linear function in two dimensions

For Problem P2

We need V to be a set of functions of Ω. In the figure on the right, we have illustrated a triangulation of a 15 sided polygonal region Ωin the plane (below), and a piecewise linear function (above, in color) of this polygon which is linear on each triangle of the triangulation; the space V would consist of functions that are linear on each triangle of the chosen triangulation.

One hopes that as the underlying triangular mesh becomes finer and finer, the solution of the discrete problem (3) will in some sense converge to the solution of the original boundary value problem P2. To measure this mesh fineness, the triangulation is indexed by a real valued parameter $h > 0$ which one takes to be very small. This parameter will be related to the size of the largest or average triangle in the triangulation. As we refine the triangulation, the space of piecewise linear functions V must also change with h. For this reason, one often reads V_h instead of V in the literature. Since we do not perform such an analysis, we will not use this notation.

Choosing a Basis

Interpolation of a Bessel Function

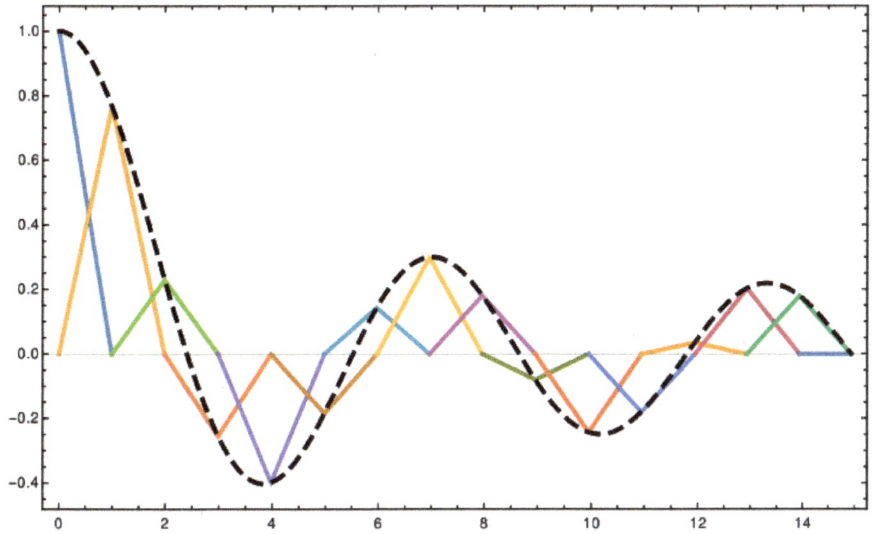

16 scaled and shifted triangular basis functions (colors) used to reconstruct a zeroeth order Bessel function J_0 (black).

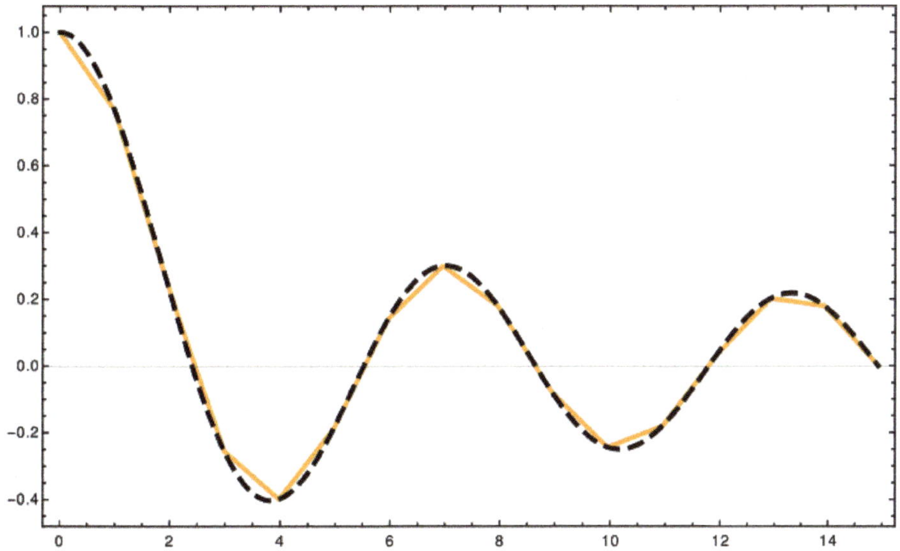

The linear combination of basis functions (yellow) reproduces J_0(blue) to any desired accuracy.

To complete the discretization, we must select a basis of V. In the one-dimensional case, for each control point x_k we will choose the piecewise linear function v_k in V whose value is 1 at x_k and zero at every $x_j,\ j \neq k$, i.e.,

$$v_k(x) = \begin{cases} \frac{x - x_{k-1}}{x_k - x_{k-1}} & \text{if } x \in [x_{k-1}, x_k], \\ \frac{x_{k+1} - x}{x_{k+1} - x_k} & \text{if } x \in [x_k, x_{k+1}], \\ 0 & \text{otherwise,} \end{cases}$$

for $k = 1, \ldots, n$; this basis is a shifted and scaled tent function. For the two-dimensional case, we choose again one basis function v_k per vertex x_k of the triangulation of the planar region Ω. The function v_k is the unique function of V whose value is 1 at x_k and zero at every $x_j,\ j \neq k$.

.

Depending on the author, the word "element" in "finite element method" refers either to the triangles in the domain, the piecewise linear basis function, or both. So for instance, an author interested in curved domains might replace the triangles with curved primitives, and so might describe the elements as being curvilinear. On the other hand, some authors replace "piecewise linear" by "piecewise quadratic" or even "piecewise polynomial". The author might then say "higher order element" instead of "higher degree polynomial". Finite element method is not restricted to triangles (or tetrahedra in 3-d, or higher order simplexes in multidimensional spaces), but can be defined on quadrilateral subdomains (hexahedra, prisms, or pyramids in 3-d, and so on). Higher order shapes (curvilinear elements) can be defined with polynomial and even non-polynomial shapes (e.g. ellipse or circle).

Examples of methods that use higher degree piecewise polynomial basis functions are the hp-FEM and spectral FEM.

More advanced implementations (adaptive finite element methods) utilize a method to assess the quality of the results (based on error estimation theory) and modify the mesh during the solution aiming to achieve approximate solution within some bounds from the 'exact' solution of the continuum problem. Mesh adaptivity may utilize various techniques, the most popular are:

- moving nodes (r-adaptivity)
- refining (and unrefining) elements (h-adaptivity)
- changing order of base functions (p-adaptivity)
- combinations of the above (hp-adaptivity).

Small Support of the Basis

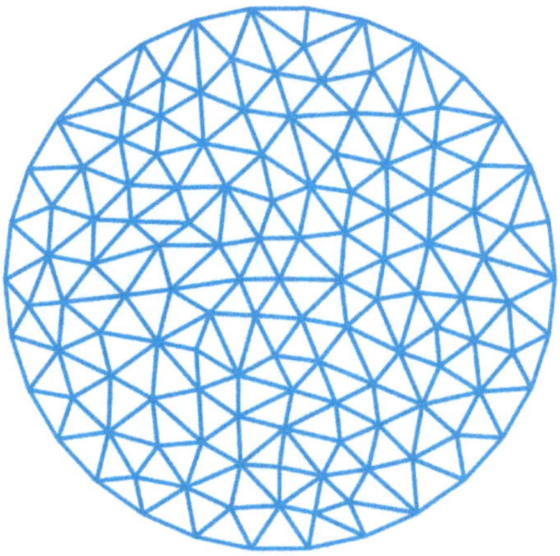

Solving the two-dimensional problem $u_{xx} + u_{yy} = -4$ in the disk centered at the origin and radius 1, with zero boundary conditions.
(a) The triangulation.

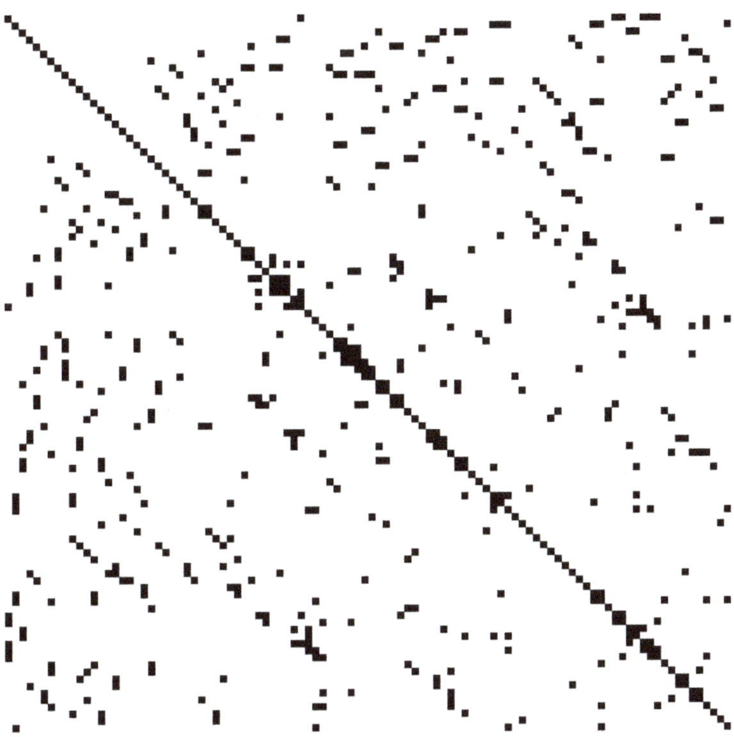

(b) The sparse matrix L of the discretized linear system

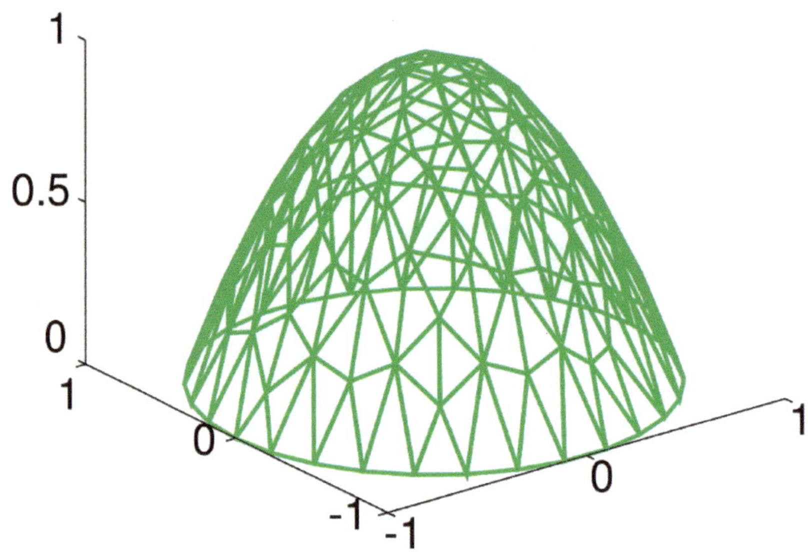

(c) The computed solution, $u(x, y) = 1 - x^2 - y^2$.

The primary advantage of this choice of basis is that the inner products

$$\langle v_j, v_k \rangle = \int_0^1 v_j v_k \, dx$$

and

$$\phi(v_j, v_k) = \int_0^1 v_j' v_k' \, dx$$

will be zero for almost all j, k. (The matrix containing $\langle v_j, v_k \rangle$ in the (j, k) location is known as the Gramian matrix.) In the one dimensional case, the support of v_k is the interval $[x_{k-1}, x_{k+1}]$. Hence, the integrands of $\langle v_j, v_k \rangle$ and $\phi(v_j, v_k)$ are identically zero whenever $|j - k| > 1$.

Similarly, in the planar case, if x_j and x_k do not share an edge of the triangulation, then the integrals

$$\int_\Omega v_j v_k \, ds$$

and

$$\int_\Omega \nabla v_j \cdot \nabla v_k \, ds$$

are both zero.

Matrix form of the Problem

If we write $u(x) = \sum_{k=1}^{n} u_k v_k(x)$ and $f(x) = \sum_{k=1}^{n} f_k v_k(x)$ then problem (3), taking $v(x) = v_j(x)$ for $j = 1, \ldots, n$, becomes

$$-\sum_{k=1}^{n} u_k \phi(v_k, v_j) = \sum_{k=1}^{n} f_k \int v_k v_j dx \text{ for } j = 1, \ldots, n. \quad (4)$$

If we denote by u and f the column vectors $(u_1, \ldots, u_n)^t$ and $(f_1, \ldots, f_n)^t$, and if we let

$$L = (L_{ij})$$

and

$$M = (M_{ij})$$

be matrices whose entries are

$$L_{ij} = \phi(v_i, v_j)$$

and

$$M_{ij} = \int v_i v_j dx$$

then we may rephrase (4) as

$$-L\mathbf{u} = M\mathbf{f}.$$

It is not necessary to assume $f(x) = \sum_{k=1}^{n} f_k v_k(x)$. For a general function $f(x)$, problem (3) with $v(x) = v_j(x)$ for $j = 1, \ldots, n$ becomes actually simpler, since no matrix M is used,

$$-L\mathbf{u} = \mathbf{b}, \quad (6)$$

where $\mathbf{b} = (b_1, \ldots, b_n)^t$ and $b_j = \int f v_j dx$ for $j = 1, \ldots, n$.

As we have discussed before, most of the entries of L and M are zero because the basis functions k have small support. So we now have to solve a linear system in the unknown v_k where most of the entries of the matrix L, which we need to invert, are zero.

Such matrices are known as sparse matrices, and there are efficient solvers for such problems (much more efficient than actually inverting the matrix.) In addition, L is symmetric and positive definite, so a technique such as the conjugate gradient method is favored. For problems that are not too large, sparse LU decompositions and Cholesky decompositions still work well. For instance, MATLAB's backslash operator (which uses sparse LU, sparse Cholesky, and other factorization methods) can be sufficient for meshes with a hundred thousand vertices.

The matrix L is usually referred to as the stiffness matrix, while the matrix M is dubbed the mass matrix.

General Form of the Finite Element Method

In general, the finite element method is characterized by the following process.

- One chooses a grid for Ω. In the preceding treatment, the grid consisted of triangles, but one can also use squares or curvilinear polygons.
- Then, one chooses basis functions. In our discussion, we used piecewise linear basis functions, but it is also common to use piecewise polynomial basis functions.

A separate consideration is the smoothness of the basis functions. For second order elliptic boundary value problems, piecewise polynomial basis function that are merely continuous suffice (i.e., the derivatives are discontinuous.) For higher order partial differential equations, one must use smoother basis functions. For instance, for a fourth order problem such as $u_{xxxx} + u_{yyyy} = f$, one may use piecewise quadratic basis functions that are C^1.

Another consideration is the relation of the finite-dimensional space V to its infinite-dimensional counterpart, in the examples above H_0^1. A conforming element method is one in which the space V is a subspace of the element space for the continuous problem. The example above is such a method. If this condition is not satisfied, we obtain a nonconforming element method, an example of which is the space of piecewise linear functions over the mesh which are continuous at each edge midpoint. Since these functions are in general discontinuous along the edges, this finite-dimensional space is not a subspace of the original H_0^1.

Typically, one has an algorithm for taking a given mesh and subdividing it. If the main method for increasing precision is to subdivide the mesh, one has an *h*-method (*h* is customarily the diameter of the largest element in the mesh.) In this manner, if one shows that the error with a grid h is bounded above by Ch^p, for some $C < \infty$ and $p > 0$, then one has an order *p* method. Under certain hypotheses (for instance, if the domain is convex), a piecewise polynomial of order d method will have an error of order $p = d + 1$.

If instead of making *h* smaller, one increases the degree of the polynomials used in the basis function, one has a *p*-method. If one combines these two refinement types, one obtains an *hp*-method (hp-FEM). In the hp-FEM, the polynomial degrees can vary from element to element. High order methods with large uniform *p* are called spectral finite element methods (SFEM). These are not to be confused with spectral methods.

For vector partial differential equations, the basis functions may take values in $\mathbb{R}^n$.

Various Types of Finite Element Methods

AEM

The Applied Element Method, or AEM combines features of both FEM and Discrete element method, or (DEM).

Generalized Finite Element Method

The generalized finite element method (GFEM) uses local spaces consisting of functions, not necessarily polynomials, that reflect the available information on the unknown solution and thus ensure good local approximation. Then a partition of unity is used to "bond" these spaces together to form the approximating subspace. The effectiveness of GFEM has been shown when applied to problems with domains having complicated boundaries, problems with micro-scales, and problems with boundary layers.

Mixed Finite Element Method

The mixed finite element method is a type of finite element method in which extra independent variables are introduced as nodal variables during the discretization of a partial differential equation problem.

hp-FEM

The hp-FEM combines adaptively, elements with variable size h and polynomial degree p in order to achieve exceptionally fast, exponential convergence rates.

hpk-FEM

The hpk-FEM combines adaptively, elements with variable size h, polynomial degree of the local approximations p and global differentiability of the local approximations $(k-1)$ in order to achieve best convergence rates.

XFEM

The extended finite element method (XFEM) is a numerical technique based on the generalized finite element method (GFEM) and the partition of unity method (PUM). It extends the classical finite element method by enriching the solution space for solutions to differential equations with discontinuous functions. Extended finite element methods enrich the approximation space so that it is able to naturally reproduce the challenging feature associated with the problem of interest: the discontinuity, singularity, boundary layer, etc. It was shown that for some problems, such an embedding of the problem's feature into the approximation space can significantly improve convergence rates and accuracy. Moreover, treating problems with discontinuities with XFEMs suppresses the need to mesh and remesh the discontinuity surfaces, thus alleviating the computational costs and projection errors associated with conventional finite element methods, at the cost of restricting the discontinuities to mesh edges.

Several research codes implement this technique to various degrees: 1. GetFEM++ 2. xfem++ 3. openxfem++

XFEM has also been implemented in codes like Altair Radioss, ASTER, Morfeo and Abaqus. It is increasingly being adopted by other commercial finite element software, with a few plugins and actual core implementations available (ANSYS, SAMCEF, OOFELIE, etc.).

S-FEM

The S-FEM, Smoothed Finite Element Methods, are a particular class of numerical simulation algorithms for the simulation of physical phenomena. It was developed by combining meshfree methods with the finite element method.

Link with the Gradient Discretisation Method

Some types of finite element methods (conforming, nonconforming, mixed finite element methods) are particular cases of the gradient discretisation method (GDM). Hence the convergence properties of the GDM, which are established for a series of problems (linear and non linear elliptic problems, linear, nonlinear and degenerate parabolic problems), hold as well for these particular finite element methods.

Comparison to the Finite Difference Method

The finite difference method (FDM) is an alternative way of approximating solutions of PDEs. The differences between FEM and FDM are:

- The most attractive feature of the FEM is its ability to handle complicated geometries (and boundaries) with relative ease. While FDM in its basic form is restricted to handle rectangular shapes and simple alterations thereof, the handling of geometries in FEM is theoretically straightforward.
- The most attractive feature of finite differences is that it can be very easy to implement.
- There are several ways one could consider the FDM a special case of the FEM approach. E.g., first order FEM is identical to FDM for Poisson's equation, if the problem is discretized by a regular rectangular mesh with each rectangle divided into two triangles.
- There are reasons to consider the mathematical foundation of the finite element approximation more sound, for instance, because the quality of the approximation between grid points is poor in FDM.
- The quality of a FEM approximation is often higher than in the corresponding FDM approach, but this is extremely problem-dependent and several examples to the contrary can be provided.

Generally, FEM is the method of choice in all types of analysis in structural mechanics (i.e. solving for deformation and stresses in solid bodies or dynamics of structures) while computational fluid dynamics (CFD) tends to use FDM or other methods like finite volume method (FVM). CFD problems usually require discretization of the problem into a large number of cells/gridpoints (millions and more), therefore cost of the solution favors simpler, lower order approximation within each cell. This is especially true for 'external flow' problems, like air flow around the car or airplane, or weather simulation.

Application

A variety of specializations under the umbrella of the mechanical engineering discipline (such as aeronautical, biomechanical, and automotive industries) commonly use integrated FEM in design and development of their products. Several modern FEM packages include specific components such as thermal, electromagnetic, fluid, and structural working environments. In a structural simulation, FEM helps tremendously in producing stiffness and strength visualizations and also in minimizing weight, materials, and costs.

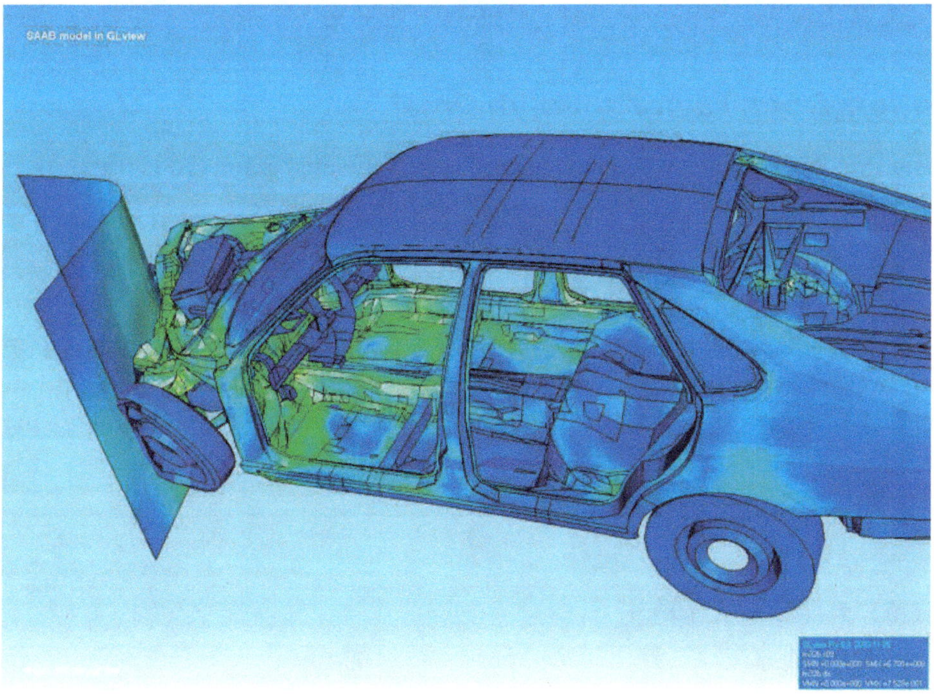

Visualization of how a car deforms in an asymmetrical crash using finite element analysis.

FEM allows detailed visualization of where structures bend or twist, and indicates the distribution of stresses and displacements. FEM software provides a wide range of simulation options for controlling the complexity of both modeling and analysis of a system. Similarly, the desired level of accuracy required and associated computational time requirements can be managed simultaneously to address most engineering applications. FEM allows entire designs to be constructed, refined, and optimized before the design is manufactured.

This powerful design tool has significantly improved both the standard of engineering designs and the methodology of the design process in many industrial applications. The introduction of FEM has substantially decreased the time to take products from concept to the production line. It is primarily through improved initial prototype designs using FEM that testing and development have been accelerated. In summary, benefits of FEM include increased accuracy, enhanced design and better insight into critical design parameters, virtual prototyping, fewer hardware prototypes, a faster and less expensive design cycle, increased productivity, and increased revenue.

FEA has also been proposed to use in stochastic modelling for numerically solving probability models.

Finite Difference Method

In mathematics, finite-difference methods (FDM) are numerical methods for solving differential equations by approximating them with difference equations, in which finite differences approximate the derivatives. FDMs are thus discretization methods.

Today, FDMs are the dominant approach to numerical solutions of partial differential equations.

Derivation from Taylor's polynomial

First, assuming the function whose derivatives are to be approximated is properly-behaved, by Taylor's theorem, we can create a Taylor Series expansion

$$f(x_0 + h) = f(x_0) + \frac{f'(x_0)}{1!}h + \frac{f^{(2)}(x_0)}{2!}h^2 + \cdots + \frac{f^{(n)}(x_0)}{n!}h^n + R_n(x),$$

where $n!$ denotes the factorial of n, and $R_n(x)$ is a remainder term, denoting the difference between the Taylor polynomial of degree n and the original function. We will derive an approximation for the first derivative of the function "f" by first truncating the Taylor polynomial:

$$f(x_0 + h) = f(x_0) + f'(x_0)h + R_1(x),$$

Setting, x_0=a we have,

$$f(a + h) = f(a) + f'(a)h + R_1(x),$$

Dividing across by h gives:

$$\frac{f(a+h)}{h} = \frac{f(a)}{h} + f'(a) + \frac{R_1(x)}{h}$$

Solving for f'(a):

$$f'(a) = \frac{f(a+h) - f(a)}{h} - \frac{R_1(x)}{h}$$

Assuming that $R_1(x)$ is sufficiently small, the approximation of the first derivative of "f" is:

$$f'(a) \approx \frac{f(a+h) - f(a)}{h}.$$

Accuracy and Order

The error in a method's solution is defined as the difference between the approximation and the exact analytical solution. The two sources of error in finite difference methods are round-off error, the loss of precision due to computer rounding of decimal quantities, and truncation error or discretization error, the difference between the exact solution of the original differential equation and the exact quantity assuming perfect arithmetic (that is, assuming no round-off).

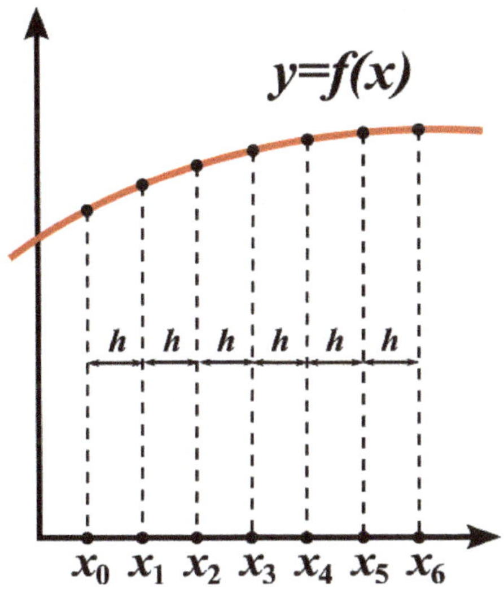

The finite difference method relies on discretizing a function on a grid.

To use a finite difference method to approximate the solution to a problem, one must first discretize the problem's domain. This is usually done by dividing the domain into a uniform grid (see image to the right). Note that this means that finite-difference methods produce sets of discrete numerical approximations to the derivative, often in a "time-stepping" manner.

An expression of general interest is the local truncation error of a method. Typically expressed using Big-O notation, local truncation error refers to the error from a single application of a method. That is, it is the quantity $f'(x_i) - f'_i$ if $f'(x_i)$ refers to the exact value and f'_i to the numerical approximation. The remainder term of a Taylor polynomial is convenient for analyzing the local truncation error. Using the Lagrange form of the remainder from the Taylor polynomial for $f(x_0 + h)$, which is

$$R_n(x_0 + h) = \frac{f^{(n+1)}(\xi)}{(n+1)!}(h)^{n+1}, \text{ where } x_0 < \xi < x_0 + h,$$

the dominant term of the local truncation error can be discovered. For example, again using the forward-difference formula for the first derivative, knowing that $f(x_i) = f(x_0 + ih)$,

$$f(x_0 + ih) = f(x_0) + f'(x_0)ih + \frac{f''(\xi)}{2!}(ih)^2,$$

and with some algebraic manipulation, this leads to

$$\frac{f(x_0 + ih) - f(x_0)}{ih} = f'(x_0) + \frac{f''(\xi)}{2!}ih,$$

and further noting that the quantity on the left is the approximation from the finite difference method and that the quantity on the right is the exact quantity of interest plus a remainder, clearly that remainder is the local truncation error. A final expression of this example and its order is:

$$\frac{f(x_0+ih)-f(x_0)}{ih}=f'(x_0)+O(h).$$

This means that, in this case, the local truncation error is proportional to the step sizes. The quality and duration of simulated FDM solution depends on the discretization equation selection and the step sizes (time and space steps). The data quality and simulation duration increase significantly with smaller step size. Therefore, a reasonable balance between data quality and simulation duration is necessary for practical usage. Large time steps are useful for increasing simulation speed in practice. However, time steps which are too large may create instabilities and affect the data quality.

The von Neumann method is usually applied to determine the numerical model stability.

Example: Ordinary Differential Equation

For example, consider the ordinary differential equation

$$u'(x)=3u(x)+2.$$

The Euler method for solving this equation uses the finite difference quotient

$$\frac{u(x+h)-u(x)}{h}\approx u'(x)$$

to approximate the differential equation by first substituting it for u'(x) then applying a little algebra (multiplying both sides by h, and then adding u(x) to both sides) to get

$$u(x+h)=u(x)+h(3u(x)+2).$$

The last equation is a finite-difference equation, and solving this equation gives an approximate solution to the differential equation.

Example: The Heat Equation

Consider the normalized heat equation in one dimension, with homogeneous Dirichlet boundary conditions

$$U_t=U_{xx}$$

$$U(0,t)=U(1,t)=0 \text{ (boundary condition)}$$

$$U(x,0)=U_0(x) \text{ (initial condition)}$$

One way to numerically solve this equation is to approximate all the derivatives by finite differences. We partition the domain in space using a mesh $x_0,\ldots,x_J$ and in time using a mesh $t_0,\ldots,t_N$. We assume a uniform partition both in space and in time, so the difference between two consecutive space points will be *h* and between two consecutive time points will be *k*. The points

$$u(x_j,t_n)=u_j^n$$

will represent the numerical approximation of $u(x_j,t_n)$

Explicit Method

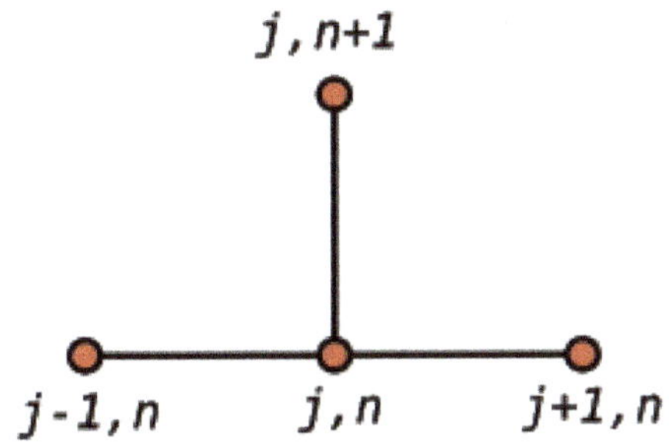

The stencil for the most common explicit method for the heat equation.

Using a forward difference at time t_n and a second-order central difference for the space derivative at position x_j (FTCS) we get the recurrence equation:

$$\frac{u_j^{n+1} - u_j^n}{k} = \frac{u_{j+1}^n - 2u_j^n + u_{j-1}^n}{h^2}.$$

This is an explicit method for solving the one-dimensional heat equation.

We can obtain u_j^{n+1} from the other values this way:

$$u_j^{n+1} = (1 - 2r)u_j^n + ru_{j-1}^n + ru_{j+1}^n$$

where $r = k/h^2$

So, with this recurrence relation, and knowing the values at time n, one can obtain the corresponding values at time n+1. u_0^n and u_J^n must be replaced by the boundary conditions, in this example they are both 0.

This explicit method is known to be numerically stable and convergent whenever $r \leq 1/2$. The numerical errors are proportional to the time step and the square of the space step:

$$\Delta u = O(k) + O(h^2)$$

Implicit Method

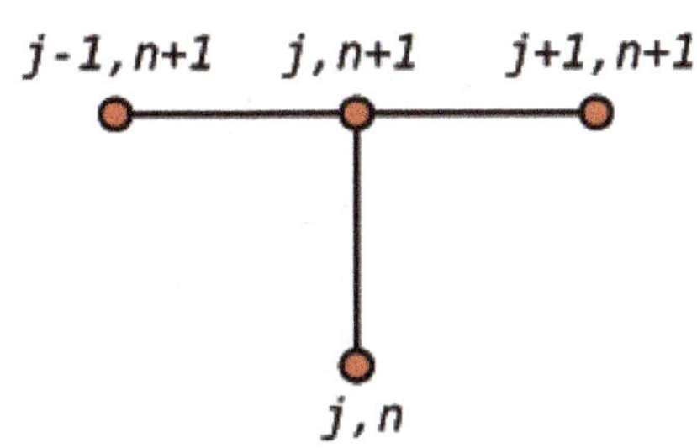

The implicit method stencil.

If we use the backward difference at time $t_{n+1/2}$ and a second-order central difference for the space derivative at position x_j (The Backward Time, Centered Space Method "BTCS") we get the recurrence equation:

$$\frac{u_j^{n+1} - u_j^n}{k} = \frac{u_{j+1}^{n+1} - 2u_j^{n+1} + u_{j-1}^{n+1}}{h^2}.$$

This is an implicit method for solving the one-dimensional heat equation.

We can obtain u_j^n from solving a system of linear equations:

$$(1+2r)u_j^{n+1} - ru_{j-1}^{n+1} - ru_{j+1}^{n+1} = u_j^n$$

The scheme is always numerically stable and convergent but usually more numerically intensive than the explicit method as it requires solving a system of numerical equations on each time step. The errors are linear over the time step and quadratic over the space step: $\Delta u = O(k) + O(h^2)$.

Crank–Nicolson Method

Finally if we use the central difference at time $t_{n+1/2}$ and a second-order central difference for the space derivative at position x_j ("CTCS") we get the recurrence equation:

$$\frac{u_j^{n+1} - u_j^n}{k} = \frac{1}{2}\left(\frac{u_{j+1}^{n+1} - 2u_j^{n+1} + u_{j-1}^{n+1}}{h^2} + \frac{u_{j+1}^n - 2u_j^n + u_{j-1}^n}{h^2}\right).$$

This formula is known as the Crank–Nicolson method.

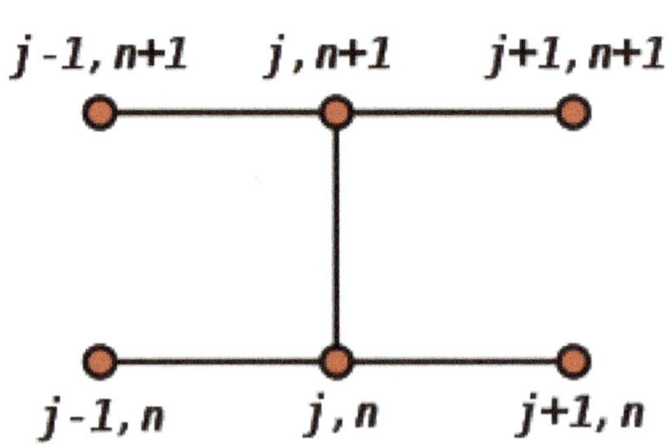

The Crank–Nicolson stencil.

We can obtain u_j^{n+1} from solving a system of linear equations:

$$(2+2r)u_j^{n+1} - ru_{j-1}^{n+1} - ru_{j+1}^{n+1} = (2-2r)u_j^n + ru_{j-1}^n + ru_{j+1}^n$$

The scheme is always numerically stable and convergent but usually more numerically intensive as it requires solving a system of numerical equations on each time step. The errors are quadratic over both the time step and the space step:

$$\Delta u = O(k^2) + O(h^2).$$

Usually the Crank–Nicolson scheme is the most accurate scheme for small time steps. The explicit scheme is the least accurate and can be unstable, but is also the easiest to implement and the least numerically intensive. The implicit scheme works the best for large time steps.

Comparison

The figures below present the solutions given by the above methods to approximate the heat equation

$$U_t = \alpha U_{xx}, \quad \alpha = \frac{1}{\pi^2},$$

with the boundary condition

$$U(0,t) = U(1,t) = 0.$$

The exact solution is

$$U(x,t) = \frac{1}{\pi^2} e^{-t} \sin(\pi x).$$

Comparison of Finite Difference Methods

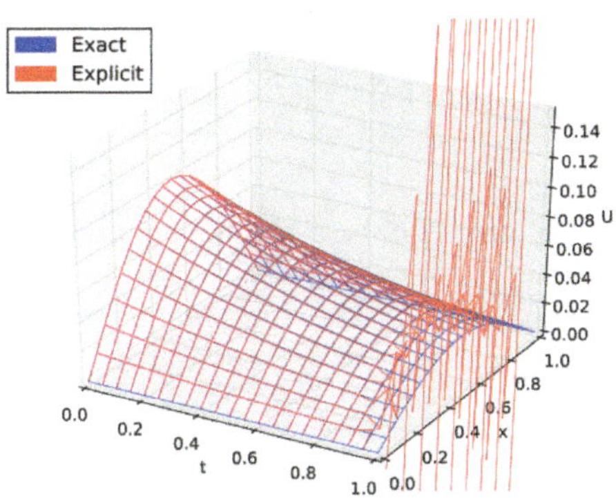

Explicit method (*not* stable)

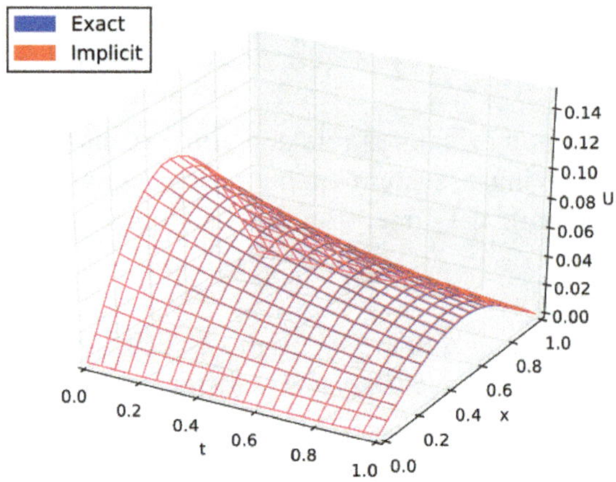

Implicit method (stable)

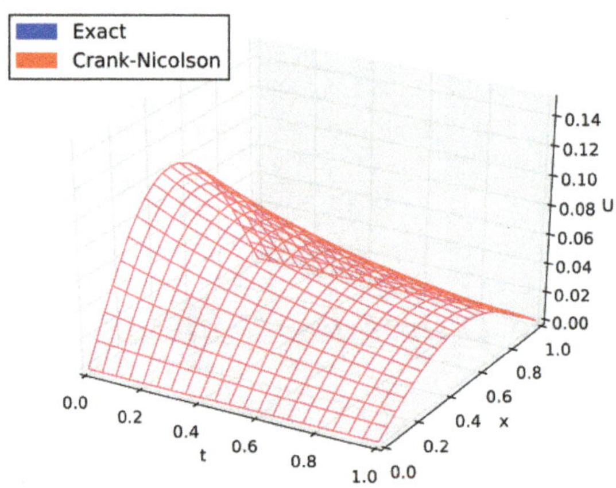

Crank-Nicolson method (stable)

Finite Volume Method

The finite volume method (FVM) is a method for representing and evaluating partial differential equations in the form of algebraic equations [LeVeque, 2002; Toro, 1999]. Similar to the finite difference method or finite element method, values are calculated at discrete places on a meshed geometry. "Finite volume" refers to the small volume surrounding each node point on a mesh. In the finite volume method, volume integrals in a partial differential equation that contain a divergence term are converted to surface integrals, using the divergence theorem. These terms are then evaluated as fluxes at the surfaces of each finite volume. Because the flux entering a given volume is identical to that leaving the adjacent volume, these methods are conservative. Another advantage of the finite volume method is that it is easily formulated to allow for unstructured meshes. The method is used in many computational fluid dynamics packages.

1D Example

Consider a simple 1D advection problem defined by the following partial differential equation

$$(1) \qquad \frac{\partial \rho}{\partial t} + \frac{\partial f}{\partial x} = 0, \quad t \geq 0.$$

Here, $\rho = \rho(x,t)$ represents the state variable and $f = f(\rho(x,t))$ represents the flux or flow of ρ. Conventionally, positive f represents flow to the right while negative f represents flow to the left. If we assume that equation (1) represents a flowing medium of constant area, we can sub-divide the spatial domain, x, into *finite volumes* or *cells* with cell centres indexed as i. For a particular cell, i, we can define the *volume average* value of

$\rho_i(t) = \rho(x,t)$ at time $t = t_1$ and $x \in \left[x_{i-\frac{1}{2}}, x_{i+\frac{1}{2}}\right]$, as

$$(2) \qquad \bar{\rho}_i(t_1) = \frac{1}{x_{i+\frac{1}{2}} - x_{i-\frac{1}{2}}} \int_{x_{i-\frac{1}{2}}}^{x_{i+\frac{1}{2}}} \rho(x,t_1)\,dx,$$

and at time $t = t_2$ as,

$$(3) \qquad \bar{\rho}_i(t_2) = \frac{1}{x_{i+\frac{1}{2}} - x_{i-\frac{1}{2}}} \int_{x_{i-\frac{1}{2}}}^{x_{i+\frac{1}{2}}} \rho(x,t_2)\,dx,$$

where $x_{i-\frac{1}{2}}$ and $x_{i+\frac{1}{2}}$ represent locations of the upstream and downstream faces or edges respectively of the ith cell.

Integrating equation (1) in time, we have:

$$(4) \qquad \rho(x,t_2) = \rho(x,t_1) - \int_{t_1}^{t_2} f_x(x,t)\,dt,$$

where $f_x = \dfrac{\partial f}{\partial x}$.

To obtain the volume average of $\rho(x,t)$ at time $t = t_2$, we integrate $\rho(x,t_2)$ over the cell volume, $\left[x_{i-\frac{1}{2}}, x_{i+\frac{1}{2}}\right]$ and divide the result by $\Delta x_i = x_{i+\frac{1}{2}} - x_{i-\frac{1}{2}}$, i.e.

$$(5) \qquad \bar{\rho}_i(t_2) = \frac{1}{\Delta x_i} \int_{x_{i-\frac{1}{2}}}^{x_{i+\frac{1}{2}}} \left\{\rho(x,t_1) - \int_{t_1}^{t_2} f_x(x,t)\,dt\right\} dx.$$

We assume that f is well behaved and that we can reverse the order of integration. Also, recall that flow is normal to the unit area of the cell. Now, since in one dimension $f_x \triangleq \nabla \cdot f$, we can apply the divergence theorem, i.e. $\oint_v \nabla \cdot f dv = \oint_S f\, dS$, and substitute for the volume integral of the divergence with the values of $f(x)$ evaluated at the cell surface (edges $x_{i-\frac{1}{2}}$ and $x_{i+\frac{1}{2}}$) of the finite volume as follows:

$$\text{(6)} \qquad \bar{\rho}_i(t_2) = \bar{\rho}_i(t_1) - \frac{1}{\Delta x_i}\left(\int_{t_1}^{t_2} f_{i+\frac{1}{2}} dt - \int_{t_1}^{t_2} f_{i-\frac{1}{2}} dt\right).$$

where $f_{i\pm\frac{1}{2}} = f\left(x_{i\pm\frac{1}{2}}, t\right)$.

We can therefore derive a *semi-discrete* numerical scheme for the above problem with cell centres indexed as i, and with cell edge fluxes indexed as $i \pm \frac{1}{2}$, by differentiating (6) with respect to time to obtain:

$$\text{(7)} \qquad \frac{d\bar{\rho}_i}{dt} + \frac{1}{\Delta x_i}\left[f_{i+\frac{1}{2}} - f_{i-\frac{1}{2}}\right] = 0,$$

where values for the edge fluxes, $f_{i\pm\frac{1}{2}}$, can be reconstructed by interpolation or extrapolation of the cell averages. Equation (7) is *exact* for the volume averages; i.e., no approximations have been made during its derivation.

This method can also be applied to a 2D situation by considering the north and south faces along with the east and west faces around a node.

General Conservation Law

We can also consider the general conservation law problem, represented by the following PDE,

$$\text{(8)} \qquad \frac{\partial \mathbf{u}}{\partial t} + \nabla \cdot \mathbf{f}(\mathbf{u}) = \mathbf{0}.$$

Here, u represents a vector of states and {\displaystyle \mathbf {f} \ } represents the corresponding flux tensor. Again we can sub-divide the spatial domain into finite volumes or cells. For a particular cell, i, we take the volume integral over the total volume of the cell, v_i, which gives,

$$\text{(9)} \qquad \int_{v_i} \frac{\partial \mathbf{u}}{\partial t} dv + \int_{v_i} \nabla \cdot \mathbf{f}(\mathbf{u})\, dv = \mathbf{0}.$$

On integrating the first term to get the *volume average* and applying the *divergence theorem* to the second, this yields

$$\text{(10)} \qquad v_i \frac{d\bar{\mathbf{u}}_i}{dt} + \oint_{S_i} \mathbf{f}(\mathbf{u}) \cdot \mathbf{n}\, dS = \mathbf{0},$$

where S_i represents the total surface area of the cell and n is a unit vector normal to the surface and pointing outward. So, finally, we are able to present the general result equivalent to (8), i.e.

$$\frac{d\bar{\mathbf{u}}_i}{dt} + \frac{1}{v_i} \oint_{S_i} \mathbf{f}(\mathbf{u}) \cdot \mathbf{n}\, dS = \mathbf{0}. \tag{11}$$

Again, values for the edge fluxes can be reconstructed by interpolation or extrapolation of the cell averages. The actual numerical scheme will depend upon problem geometry and mesh construction. MUSCL reconstruction is often used in high resolution schemes where shocks or discontinuities are present in the solution.

Finite volume schemes are conservative as cell averages change through the edge fluxes. In other words, *one cell's loss is another cell's gain*!

PARTIAL DIFFERENTIAL ALGEBRAIC EQUATION

In mathematics a partial differential algebraic equation (PDAE) set is an incomplete system of partial differential equations that is closed with a set of algebraic equations.

Definition

A general PDAE is defined as:

$$0 = \mathbf{F}\left(\mathbf{x}, \mathbf{y}, \frac{\partial y_i}{\partial x_j}, \frac{\partial^2 y_i}{\partial x_j \partial x_k}, \ldots, \mathbf{z}\right),$$

where:

- $\mathbf{F}$ is a set of arbitrary functions;
- $\mathbf{x}$ is a set of independent variables;
- $\mathbf{y}$ is a set of dependent variables for which partial derivatives are defined; and
- $\mathbf{z}$ is a set of dependent variables for which no partial derivatives are defined.

The relationship between a PDAE and a partial differential equation (PDE) is analogous to the relationship between an ordinary differential equation (ODE) and a differential algebraic equation (DAE).

PDAEs of this general form are challenging to solve. Simplified forms are studied in more detail in the literature. Even as recently as 2000, the term "PDAE" has been handled as unfamiliar by those in related fields.

Solution Methods

Semi-discretization is a common method for solving PDAEs whose independent variables are those of time and space, and has been used for decades. This method involves removing the spatial variables using a discretization method, such as the finite volume method, and incorporating the resulting linear equations as part of the algebraic relations. This reduces the system to a DAE, for which conventional solution methods can be employed.

NONLINEAR PARTIAL DIFFERENTIAL EQUATION

In mathematics and physics, a **nonlinear partial differential equation** is a partial differential equation with nonlinear terms. They describe many different physical systems, ranging from gravitation to fluid dynamics, and have been used in mathematics to solve problems such as the Poincaré conjecture and the Calabi conjecture. They are difficult to study: there are almost no general techniques that work for all such equations, and usually each individual equation has to be studied as a separate problem.

Methods for Studying Nonlinear Partial Differential Equations

Existence and Uniqueness of Solutions

A fundamental question for any PDE is the existence and uniqueness of a solution for given boundary conditions. For nonlinear equations these questions are in general very hard: for example, the hardest part of Yau's solution of the Calabi conjecture was the proof of existence for a Monge–Ampere equation.

Singularities

The basic questions about singularities (their formation, propagation, and removal, and regularity of solutions) are the same as for linear PDE, but as usual much harder to study. In the linear case one can just use spaces of distributions, but nonlinear PDEs are not usually defined on arbitrary distributions, so one replaces spaces of distributions by refinements such as Sobolev spaces.

An example of singularity formation is given by the Ricci flow: Richard S. Hamilton showed that while short time solutions exist, singularities will usually form after a finite time. Grigori Perelman's solution of the Poincaré conjecture depended on a deep study of these singularities, where he showed how to continue the solution past the singularities.

Linear Approximation

The solutions in a neighborhood of a known solution can sometimes be studied by linearizing the PDE around the solution. This corresponds to studying the tangent space of a point of the moduli space of all solutions.

Moduli Space of Solutions

Ideally one would like to describe the (moduli) space of all solutions explicitly, and for some very special PDEs this is possible. (In general this is a hopeless problem:

it is unlikely that there is any useful description of all solutions of the Navier–Stokes equation for example, as this would involve describing all possible fluid motions.) If the equation has a very large symmetry group, then one is usually only interested in the moduli space of solutions modulo the symmetry group, and this is sometimes a finite-dimensional compact manifold, possibly with singularities; for example, this happens in the case of the Seiberg–Witten equations. A slightly more complicated case is the self dual Yang–Mills equations, when the moduli space is finite-dimensional but not necessarily compact, though it can often be compactified explicitly. Another case when one can sometimes hope to describe all solutions is the case of completely integrable models, when solutions are sometimes a sort of superposition of solitons; for example, this happens for the Korteweg–de Vries equation.

Exact Solutions

It is often possible to write down some special solutions explicitly in terms of elementary functions (though it is rarely possible to describe all solutions like this). One way of finding such explicit solutions is to reduce the equations to equations of lower dimension, preferably ordinary differential equations, which can often be solved exactly. This can sometimes be done using separation of variables, or by looking for highly symmetric solutions.

Some equations have several different exact solutions.

Numerical Solutions

Numerical solution on a computer is almost the only method that can be used for getting information about arbitrary systems of PDEs. There has been a lot of work done, but a lot of work still remains on solving certain systems numerically, especially for the Navier–Stokes and other equations related to weather prediction.

Lax Pair

If a system of PDEs can be put into Lax pair form

$$\frac{dL}{dt} = LA - AL$$

then it usually has an infinite number of first integrals, which help to study it.

Euler–Lagrange Equations

Systems of PDEs often arise as the Euler–Lagrange equations for a variational problem. Systems of this form can sometimes be solved by finding an extremum of the original variational problem.

Integrable Systems

PDEs that arise from integrable systems are often the easiest to study, and can sometimes be completely solved. A well-known example is the Korteweg–de Vries equation.

Symmetry

Some systems of PDEs have large symmetry groups. For example, the Yang–Mills equations are invariant under an infinite-dimensional gauge group, and many systems of equations (such as the Einstein field equations) are invariant under diffeomorphisms of the underlying manifold. Any such symmetry groups can usually be used to help study the equations; in particular if one solution is known one can trivially generate more by acting with the symmetry group.

Sometimes equations are parabolic or hyperbolic "modulo the action of some group": for example, the Ricci flow equation is not quite parabolic, but is "parabolic modulo the action of the diffeomorphism group", which implies that it has most of the good properties of parabolic equations.

REFERENCES

1. Adomian, G. (1994). Solving Frontier problems of Physics: The decomposition method. Kluwer Academic Publishers.
2. Courant, R. & Hilbert, D. (1962), Methods of Mathematical Physics, II, New York: Wiley-Interscience.
3. Evans, L. C. (1998), Partial Differential Equations, Providence: American Mathematical Society, ISBN 0-8218-0772-2.
4. Holubová, Pavel Drábek ; Gabriela (2007). Elements of partial differential equations ([Online-Ausg.]. ed.). Berlin: de Gruyter. ISBN 9783110191240.
5. Ibragimov, Nail H (1993), CRC Handbook of Lie Group Analysis of Differential Equations Vol. 1-3, Providence: CRC-Press, ISBN 0-8493-4488-3.
6. John, F. (1982), Partial Differential Equations (4th ed.), New York: Springer-Verlag, ISBN 0-387-90609-6.
7. Jost, J. (2002), Partial Differential Equations, New York: Springer-Verlag, ISBN 0-387-95428-7.
8. Lewy, Hans (1957), "An example of a smooth linear partial differential equation without solution", Annals of Mathematics. Second Series, 66 (1): 155–158, doi:10.2307/1970121.
9. Liao, S.J. (2003), Beyond Perturbation: Introduction to the Homotopy Analysis Method, Boca Raton: Chapman & Hall/ CRC Press, ISBN 1-58488-407-X
10. Olver, P.J. (1995), Equivalence, Invariants and Symmetry, Cambridge Press.
11. Petrovskii, I. G. (1967), Partial Differential Equations, Philadelphia: W. B.

Saunders Co..

12. Pinchover, Y. & Rubinstein, J. (2005), An Introduction to Partial Differential Equations, New York: Cambridge University Press, ISBN 0-521-84886-5.
13. Polyanin, A. D. (2002), Handbook of Linear Partial Differential Equations for Engineers and Scientists, Boca Raton: Chapman & Hall/CRC Press, ISBN 1-58488-299-9.
14. Polyanin, A. D. & Zaitsev, V. F. (2004), Handbook of Nonlinear Partial Differential Equations, Boca Raton: Chapman & Hall/CRC Press, ISBN 1-58488-355-3.
15. Polyanin, A. D.; Zaitsev, V. F. & Moussiaux, A. (2002), Handbook of First Order Partial Differential Equations, London: Taylor & Francis, ISBN 0-415-27267-X.
16. Roubíček, T. (2013), Nonlinear Partial Differential Equations with Applications (2nd ed.), Basel, Boston, Berlin: Birkhäuser, ISBN 978-3-0348-0512-4, MR MR3014456
17. Solin, P. (2005), Partial Differential Equations and the Finite Element Method, Hoboken, NJ: J. Wiley & Sons, ISBN 0-471-72070-4.
18. Solin, P.; Segeth, K. & Dolezel, I. (2003), Higher-Order Finite Element Methods, Boca Raton: Chapman & Hall/CRC Press, ISBN 1-58488-438-X.
19. Stephani, H. (1989), Differential Equations: Their Solution Using Symmetries. Edited by M. MacCallum, Cambridge University Press.
20. Wazwaz, Abdul-Majid (2009). Partial Differential Equations and Solitary Waves Theory. Higher Education Press. ISBN 978-3-642-00251-9.
21. Wazwaz, Abdul-Majid (2002). Partial Differential Equations Methods and Applications. A.A. Balkema. ISBN 90-5809-369-7.
22. Zwillinger, D. (1997), Handbook of Differential Equations (3rd ed.), Boston: Academic Press, ISBN 0-12-784395-7.
23. Gershenfeld, N. (1999), The Nature of Mathematical Modeling (1st ed.), New York: Cambridge University Press, New York, NY, USA, ISBN 0-521-57095-6.
24. Krasil'shchik, I.S. & Vinogradov, A.M., Eds. (1999), Symmetries and Conserwation Laws for Differential Equations of Mathematical Physics, American Mathematical Society, Providence, Rhode Island,USA, ISBN 0-8218-0958-X.
25. Krasil'shchik, I.S.; Lychagin, V.V. & Vinogradov, A.M. (1986), Geometry of Jet Spaces and Nonlinear Partial Differential Equations, Gordon and Breach Science Publishers, New York, London, Paris, Montreux, Tokyo, ISBN 2-88124-051-8.
26. Vinogradov, A.M. (2001), Cohomological Analysis of Partial Differential Equations and Secondary Calculus, American Mathematical Society, Providence, Rhode Island,USA, ISBN 0-8218-2922-X.
27. R. Courant and D. Hilbert, Methods of Mathematical Physics, Vol II, Wiley (Interscience), New York, 1962.
28. L.C. Evans, Partial Differential Equations, American Mathematical Society, Providence, 1998. ISBN 0-8218-0772-2

29. A. D. Polyanin, V. F. Zaitsev, and A. Moussiaux, Handbook of First Order Partial Differential Equations, Taylor & Francis, London, 2002. ISBN 0-415-27267-X
30. D. Polyanin, Handbook of Linear Partial Differential Equations for Engineers and Scientists, Chapman & Hall/CRC Press, Boca Raton, 2002. ISBN 1-58488-299-9
31. Sarra, Scott The Method of Characteristics with applications to Conservation Laws, Journal of Online Mathematics and its Applications, 2003.
32. Courant, Richard; Hilbert, David (1962), Methods of Mathematical Physics, Volume II, Wiley-Interscience
33. Delgado, Manuel (1997), "The Lagrange-Charpit Method", SIAM Review, 39 (2): 298–304, Bibcode:1997SIAMR..39..298D, doi:10.1137/S0036144595293534, JSTOR 2133111
34. Evans, Lawrence C. (1998), Partial Differential Equations, Providence: American Mathematical Society, ISBN 0-8218-0772-2
35. John, Fritz (1991), Partial differential equations (4th ed.), Springer, ISBN 978-0-387-90609-6
36. Polyanin, A. D.; Zaitsev, V. F.; Moussiaux, A. (2002), Handbook of First Order Partial Differential Equations, London: Taylor & Francis, ISBN 0-415-27267-X
37. Polyanin, A. D. (2002), Handbook of Linear Partial Differential Equations for Engineers and Scientists, Boca Raton: Chapman & Hall/CRC Press, ISBN 1-58488-299-9
38. Sarra, Scott (2003), "The Method of Characteristics with applications to Conservation Laws", Journal of Online Mathematics and its Applications.
39. Streeter, VL; Wylie, EB (1998), Fluid mechanics (International 9th Revised ed.), McGraw-Hill Higher Education
40. Reddy, J.N. (2006). An Introduction to the Finite Element Method (Third ed.). McGraw-Hill. ISBN 9780071267618.
41. Hrennikoff, Alexander (1941). "Solution of problems of elasticity by the framework method". Journal of applied mechanics. 8.4: 169–175.
42. Courant, R. (1943). "Variational methods for the solution of problems of equilibrium and vibrations". Bulletin of the American Mathematical Society. 49: 1–23. doi:10.1090/s0002-9904-1943-07818-4.
43. Hinton, Ernest; Irons, Bruce (July 1968). "Least squares smoothing of experimental data using finite elements". Strain. 4: 24–27. doi:10.1111/j.1475-1305.1968.tb01368.x.
44. "SAP-IV Software and Manuals". NISEE e-Library, The Earthquake Engineering Online Archive.
45. Gard Paulsen; Håkon With Andersen; John Petter Collett; Iver Tangen Stensrud (2014). Building Trust, The history of DNV 1864-2014. Lysaker, Norway: Dinamo Forlag A/S. pp. 121, 436. ISBN 978-82-8071-256-1.
46. Strang, Gilbert; Fix, George (1973). An Analysis of The Finite Element Method. Prentice Hall. ISBN 0-13-032946-0.

47. Zienkiewicz, O.C.; Taylor, R.L.; Zhu, J.Z. (2005). The Finite Element Method: Its Basis and Fundamentals (Sixth ed.). Butterworth-Heinemann. ISBN 0750663200.

48. Bathe, K.J. (2006). Finite Element Procedures. Cambridge, MA: Klaus-Jürgen Bathe. ISBN 097900490X.

49. Babuška, Ivo; Banerjee, Uday; Osborn, John E. (June 2004). "Generalized Finite Element Methods: Main Ideas, Results, and Perspective". International Journal of Computational Methods. 1 (1): 67–103. doi:10.1142/S0219876204000083.

50. P. Solin, K. Segeth, I. Dolezel: Higher-Order Finite Element Methods, Chapman & Hall/CRC Press, 2003

51. Hastings, J. K., Juds, M. A., Brauer, J. R., Accuracy and Economy of Finite Element Magnetic Analysis, 33rd Annual National Relay Conference, April 1985.

52. McLaren-Mercedes (2006). "McLaren Mercedes: Feature - Stress to impress". Archived from the original on 2006-10-30. Retrieved 2006-10-03.

53. Peng Long; Wang Jinliang; Zhu Qiding (19 May 1995). "Methods with high accuracy for finite element probability computing". Journal of Computational and Applied Mathematics. 59 (2): 181–189. doi:10.1016/0377-0427(94)00027-X.

54. Haldar, Achintya; Mahadevan, Sankaran (2000). Reliability Assessment Using Stochastic Finite Element Analysis. John Wiley & Sons. ISBN 978-0471369615.

55. Eymard, R. Gallouët, T. R. Herbin, R. (2000) The finite volume method Handbook of Numerical Analysis, Vol. VII, 2000, p. 713–1020. Editors: P.G. Ciarlet and J.L. Lions.

56. LeVeque, Randall (2002), Finite Volume Methods for Hyperbolic Problems, Cambridge University Press.

57. Toro, E. F. (1999), Riemann Solvers and Numerical Methods for Fluid Dynamics, Springer-Verlag.

CHAPTER 4

Differential Algebraic Equations

Chapter Contents

- **Introduction**
- **Other Forms of Daes**
- **Numerical Treatment of DAE and Applications**
- **Integro-Differential Equation**
- **Laplace Transform**
- **Fractional Calculus**

A differential-algebraic equation (DAE) is a type of differential equation in which the derivatives are not (in general) expressed explicitly, and typically derivatives of some of the dependent variables may not appear in the equations at all. The general form of a system of DAEs is given by

$F(t,x,x') = 0$

where $x' = dx/dt$.

A system of DAEs can be converted to a system of ordinary differential equations by differentiating it with respect to the independent variablet.

INTRODUCTION

In mathematics, differential-algebraic equations (DAEs) are a general form of (systems of) differential equations for vector–valued functions x in one independent variable t,

$$F(\dot{x}(t), x(t), t) = 0$$

where $x : [a, b] \to \mathbb{R}^n$ is a vector of dependent variables $x(t) = (x_1(t), \ldots, x_n(t))$ and the system has as many equations, $F = (F_1, \ldots, F_n) : \mathbb{R}^{2n+1} \to \mathbb{R}^n$. They are distinct from ordinary differential equation (ODE) in that a DAE is not completely solvable for the derivatives of all components of the function x because these may not all appear (i.e. some equations are algebraic); technically the distinction between an implicit ODE system [that may be rendered explicit] and a DAE system is that the Jacobian matrix $\frac{\partial F(u,v,t)}{\partial u}$ is a singular matrix for a DAE system. This distinction between ODEs and DAEs is made because DAEs have different characteristics and are generally more difficult to solve.

In practical terms, the distinction between DAEs and ODEs is often that the solution of a DAE system depends on the derivatives of the input signal and not just the signal itself as in the case of ODEs; this issue is commonly encountered in systems with hysteresis, such as the Schmitt trigger.

This difference is more clearly visible if the system may be rewritten so that instead of x we consider a pair xy of vectors of dependent variables and the DAE has the form

$$\begin{aligned}\dot{x}(t) &= f(x(t), y(t), t),\\ 0 &= g(x(t), y(t), t).\end{aligned}$$

where $x(t) \in \mathbb{R}^n$, $y(t) \in \mathbb{R}^m$, $f : \mathbb{R}^{n+m+1} \to \mathbb{R}^n$ and $g : \mathbb{R}^{n+m+1} \to \mathbb{R}^m$.

A DAE system of this form is called semi-explicit. Every solution of the second half g of the equation defines a unique direction for x via the first half f of the equations, while the direction for y is arbitrary. But not every point (x,y,t) is a solution of g. The variables in x and the first half f of the equations get the attribute differential. The components of y and the second half g of the equations are called the algebraic variables or equations of the system. [The term algebraic in the context of DAEs only means free of derivatives and is not related to (abstract) algebra.]

The solution of a DAE consists of two parts, first the search for consistent initial values and second the computation of a trajectory. To find consistent initial values it is often necessary to consider the derivatives of some of the component functions of the DAE. The highest order of a derivative that is necessary for this process is called the differentiation index. The equations derived in computing the index and consistent initial values may also be of use in the computation of the trajectory. A semi-explicit DAE system can be converted to an implicit one by decreasing the differentiation index by one, and vice versa.

OTHER FORMS OF DAES

The distinction of DAEs to ODEs becomes apparent if some of the dependent variables occur without their derivatives. The vector of dependent variables may then be written as pair xy and the system of differential equations of the DAE appears in the form

$$F\left(\dot{x}, x, y, t\right) = 0$$

where

x, a vector in $\mathbb{R}^n$, are dependent variables for which derivatives are present (*differential variables*),
y, a vector in $\mathbb{R}^m$, are dependent variables for which no derivatives are present (*algebraic variables*),
t, a scalar (usually time) is an independent variable.
F is a vector of $n+m$ functions that involve subsets these $n+m$ variables and n derivatives.

As a whole, the set of DAEs is a function

$F : \mathbb{R}^{(2n+m+1)} \to \mathbb{R}^{(n+m)}$.

Initial conditions must be a solution of the system of equations of the form

$F(\dot{x}(t_0),\ x(t_0), y(t_0), t_0) = 0.$

Examples

The pendulum in Cartesian coordinates (x,y) with center in (0,0) and length L has the Euler–Lagrange equations

$$\begin{aligned} \dot{x} &= u, & \dot{y} &= v, \\ \dot{u} &= \lambda x, & \dot{v} &= \lambda y - g, \\ x^2 + y^2 &= L^2, \end{aligned}$$

where λ is a Lagrange multiplier. The momentum variables u and v should be constrained by the law of conservation of energy and their direction should point along the circle. Neither condition is explicit in those equations. Differentiation of the last equation leads to

$$\begin{aligned} & \dot{x}\,x + \dot{y}\,y = 0 \\ \Rightarrow \quad & u\,x + v\,y = 0, \end{aligned}$$

restricting the direction of motion to the tangent of the circle. The next derivative of this equation implies

$$\begin{aligned} \dot{u}\,x + \dot{v}\,y + u\,\dot{x} + v\,\dot{y} &= 0, \\ \Rightarrow \quad \lambda(x^2 + y^2) - gy + u^2 + v^2 &= 0, \\ \Rightarrow \quad L^2\,\lambda - gy + u^2 + v^2 &= 0, \end{aligned}$$

and the derivative of that last identity simplifies to $L^2\dot{\lambda} - 3gv = 0$ which implicitly implies the conservation of energy since after integration the constant $E = \tfrac{3}{2}gy - \tfrac{1}{2}L^2\lambda = \frac{1}{2}(u^2 + v^2) + gy$ is the sum of kinetic and potential energy.

To obtain unique derivative values for all dependent variables the last equation was three times differentiated. This gives a differentiation index of 3, which is typical for constrained mechanical systems.

If initial values (x_0, u_0) and a sign for *y* are given, the other variables are determined via $y = \pm\sqrt{L^2 - x^2}$, and if $y \neq 0$ then $v = -ux/y$ and $\lambda = (gy - u^2 - v^2)/L^2$. To proceed to the next point it is sufficient to get the derivatives of *x* and *u*, that is, the system to solve is now

$$\begin{aligned}\dot{x} &= u,\\ \dot{u} &= \lambda x,\\ 0 &= x^2 + y^2 - L^2,\\ 0 &= ux + vy,\\ 0 &= u^2 - gy + v^2 + L^2\,\lambda.\end{aligned}$$

DAEs also naturally occur in the modelling of circuits with non-linear devices. Modified nodal analysis employing DAEs is used for example in the ubiquitous SPICE family of numeric circuit simulators. Similarly, Fraunhofer's Analog Insydes Mathematica package can be used to derive DAEs from a netlist and then simplify or even solve the equations symbolically in some cases. It is worth noting that the index of a DAE (of a circuit) can be made arbitrarily high by cascading/coupling via capacitors operational amplifiers with positive feedback.

Semi-explicit DAE of index 1

DAE of the form

$$\begin{aligned}\dot{x} &= f(x, y, t),\\ 0 &= g(x, y, t).\end{aligned}$$

are called semi-explicit. The index-1 property requires that g is solvable for y. In other words, the differentiation index is 1 if by differentiation of the algebraic equations for t an implicit ODE system results,

$$\begin{aligned}\dot{x} &= f(x, y, t)\\ 0 &= \partial_x g(x, y, t)\dot{x} + \partial_y g(x, y, t)\dot{y} + \partial_t g(x, y, t),\end{aligned}$$

which is solvable for $(\dot{x},\, \dot{y})$ if $\det\left(\partial_y g(x, y, t)\right) \neq 0$.

Every sufficiently smooth DAE is almost everywhere reducible to this semi-explicit index-1 form.

NUMERICAL TREATMENT OF DAE AND APPLICATIONS

Two major problems in the solution of DAEs are index reduction and consistent initial conditions. Most numerical solvers require ordinary differential equations and algebraic equations of the form

$$\begin{aligned}\frac{dx}{dt} &= f\left(x, y, t\right),\\ 0 &= g\left(x, y, t\right).\end{aligned}$$

It is a non-trivial task to convert arbitrary DAE systems into ODEs for solution by pure ODE solvers. Techniques which can be employed include Pantelides algorithm and dummy derivative index reduction method. Alternatively, a direct solution of high-index DAEs with inconsistent initial conditions is also possible. This solution approach involves a transformation of the derivative elements through orthogonal collocation on finite elements or direct transcription into algebraic expressions. This allows DAEs of any index to be solved without rearrangement in the open equation form

$$0 = f\left(\frac{dx}{dt}, x, y, t\right),$$
$$0 = g(x, y, t).$$

Once the model has been converted to algebraic equation form, it is solvable by large-scale nonlinear programming solvers.

Tractability

Several measures of DAEs tractability in terms of numerical methods have developed, such as differentiation index, perturbation index, tractability index, geometric index, and the Kronecker index.

Structural Analysis for DAEs

We use the Σ-method to analyze a DAE. We construct for the DAE a signature matrix $\Sigma = (\sigma_{i,j})$, where each row corresponds to each equation f_i and each column corresponds to each variable x_j. The entry in position (i, j) is $\sigma_{i,j}$, which denotes the highest order of derivative to which x_j occurs in f_i, or $-\infty$ if x_j does not occur in f_i.

For the pendulum DAE above, the variables are $(x_1, x_2, x_3, x_4, x_5) = (x, y, u, v, \lambda)$. The corresponding signature matrix is

$$\Sigma = \begin{bmatrix} 1 & - & 0^\bullet & - & - \\ - & 1^\bullet & - & 0 & - \\ 0 & - & 1 & - & 0^\bullet \\ - & 0 & - & 1^\bullet & 0 \\ 0^\bullet & 0 & - & - & - \end{bmatrix}$$

INTEGRO-DIFFERENTIAL EQUATION

In mathematics, an integro-differential equation is an equation that involves both integrals and derivatives of a function.

General first order linear equations

The general first-order, linear (only with respect to the term involving derivative) integro-differential equation is of the form

$$\frac{d}{dx}u(x) + \int_{x_0}^{x} f(t, u(t))\,dt = g(x, u(x)), \qquad u(x_0) = u_0, \qquad x_0 \geq 0.$$

As is typical with differential equations, obtaining a closed-form solution can often be difficult. In the relatively few cases where a solution can be found, it is often by some kind of integral transform, where the problem is first transformed into an algebraic setting. In such situations, the solution of the problem may be derived by applying the inverse transform to the solution of this algebraic equation.

Consider the following first-order problem,

$$u'(x) + 2u(x) + 5\int_0^x u(t)\,dt = \begin{cases} 1, & x \geq 0 \\ 0, & x < 0 \end{cases} \qquad \text{with} \qquad u(0) = 0.$$

The Laplace transform is defined by,

$$U(s) = \mathcal{L}\{u(x)\} = \int_0^\infty e^{-sx} u(x)\,dx.$$

Upon taking term-by-term Laplace transforms, and utilising the rules for derivatives and integrals, the integro-differential equation is converted into the following algebraic equation,

$$sU(s) - u(0) + 2U(s) + \frac{5}{s}U(s) = \frac{1}{s}.$$

Thus,

$$U(s) = \frac{1}{s^2 + 2s + 5}.$$

Inverting the Laplace transform using contour integral methods then gives

$$u(x) = \frac{1}{2}e^{-x}\sin(2x).$$

Applications

Integro-differential equations model many situations from science and engineering. A particularly rich source is electrical circuit analysis.

The activity of interacting inhibitory and excitatory neurons can be described by a system of integro-differential equations, see for example the Wilson-Cowan model.

LAPLACE TRANSFORM

In mathematics the Laplace transform is an integral transform named after its discoverer Pierre-Simon Laplace. It takes a function of a positive real variable t (often time) to a function of a complex variable s (frequency).

The Laplace transform is very similar to the Fourier transform. While the Fourier transform of a function is a complex function of a real variable (frequency), the Laplace transform of a function is a complex function of a complex variable. Laplace transforms are usually restricted to functions of t with t > 0. A consequence of this restriction is that the Laplace transform of a function is a holomorphic function of the variable s. Unlike the Fourier transform, the Laplace transform of a distribution is generally a well-behaved function. Also techniques of complex variables can be used directly to study Laplace transforms. As a holomorphic function, the Laplace transform has a power series representation. This power series expresses a function as a linear superposition of moments of the function. This perspective has applications in probability theory.

The Laplace transform is invertible on a large class of functions. The inverse Laplace transform takes a function of a complex variable s (often frequency) and yields a function of a real variable t (time). Given a simple mathematical or functional description of an input or output to a system, the Laplace transform provides an alternative functional description that often simplifies the process of analyzing the behavior of the system, or in synthesizing a new system based on a set of specifications. So, for example, Laplace transformation from the time domain to the frequency domain transforms differential equations into algebraic equations and convolution into multiplication. It has many applications in the sciences and technology.

History

The Laplace transform is named after mathematician and astronomer Pierre-Simon Laplace, who used a similar transform (now called the z-transform) in his work on probability theory.The current widespread use of the transform (mainly in engineering) came about during and soon after World War II although it had been used in the 19th century by Abel, Lerch, Heaviside, and Bromwich.

The early history of methods having some similarity to Laplace transform is as follows. From 1744, Leonhard Euler investigated integrals of the form

$$z = \int X(x)e^{ax}\,dx \quad \text{and} \quad z = \int X(x)x^A\,dx$$

as solutions of differential equations but did not pursue the matter very far.

Joseph Louis Lagrange was an admirer of Euler and, in his work on integrating probability density functions, investigated expressions of the form

$$\int X(x)e^{-ax}a^x\,dx,$$

which some modern historians have interpreted within modern Laplace transform theory.

These types of integrals seem first to have attracted Laplace's attention in 1782 where he was following in the spirit of Euler in using the integrals themselves as solutions of equations. However, in 1785, Laplace took the critical step forward when, rather than just looking for a solution in the form of an integral, he started to apply the transforms in the sense that was later to become popular. He used an integral of the form

$$\int x^s\phi(x)\,dx,$$

akin to a Mellin transform, to transform the whole of a difference equation, in order to look for solutions of the transformed equation. He then went on to apply the Laplace transform in the same way and started to derive some of its properties, beginning to appreciate its potential power.

Laplace also recognised that Joseph Fourier's method of Fourier series for solving the diffusion equation could only apply to a limited region of space because those solutions were periodic. In 1809, Laplace applied his transform to find solutions that diffused indefinitely in space.

Formal Definition

The Laplace transform is a frequency-domain approach for continuous time signals irrespective of whether the system is stable or unstable. The Laplace transform of a function f(t), defined for all real numbers $t \geq 0$, is the function F(s), which is a unilateral transform defined by

$$F(s) = \int_0^\infty f(t)e^{-st}\,dt$$

where s is a complex number frequency parameter

$s = \sigma + i\omega$, with real numbers σ and ω.

Other notations for the Laplace transform include L{f} , or alternatively L{f(t)} instead of F.

The meaning of the integral depends on types of functions of interest. A necessary condition for existence of the integral is that f must be locally integrable on [0, ∞). For locally integrable functions that decay at infinity or are of exponential type, the integral can be understood to be a (proper) Lebesgue integral. However, for many applications it is necessary to regard it to be a conditionally convergent improper integral at ∞. Still more generally, the integral can be understood in a weak sense, and this is dealt with below.

One can define the Laplace transform of a finite Borel measure μ by the Lebesgue integral

$$\mathcal{L}\{\mu\}(s) = \int_{[0,\infty)} e^{-st}\,d\mu(t).$$

An important special case is where μ is a probability measure, for example, the Dirac delta function. In operational calculus, the Laplace transform of a measure is often treated as though the measure came from a probability density function f. In that case, to avoid potential confusion, one often writes

$$\mathcal{L}\{f\}(s) = \int_{0^-}^\infty f(t)e^{-st}\,dt,$$

where the lower limit of 0^- is shorthand notation for

$$\lim_{\varepsilon\downarrow 0}\int_{-\varepsilon}^\infty.$$

This limit emphasizes that any point mass located at 0 is entirely captured by the Laplace transform. Although with the Lebesgue integral, it is not necessary to take such a limit, it does appear more naturally in connection with the Laplace–Stieltjes transform.

Probability Theory

In pure and applied probability, the Laplace transform is defined as an expected value. If X is a random variable with probability density function f, then the Laplace transform of f is given by the expectation

$$\mathcal{L}\{f\}(s) = E\left[e^{-sX}\right].$$

By abuse of language, this is referred to as the Laplace transform of the random variable X itself. Replacing s by −t gives the moment generating function of X. The Laplace transform has applications throughout probability theory, including first passage times of stochastic processes such as Markov chains, and renewal theory.

Of particular use is the ability to recover the cumulative distribution function of a continuous random variable X by means of the Laplace transform as follows

$$F_X(x) = \mathcal{L}^{-1}\left\{\frac{1}{s}E\left[e^{-sX}\right]\right\}(x) = \mathcal{L}^{-1}\left\{\frac{1}{s}\mathcal{L}\{f\}(s)\right\}(x).$$

Bilateral Laplace Transform

When one says “the Laplace transform” without qualification, the unilateral or one-sided transform is normally intended. The Laplace transform can be alternatively defined as the bilateral Laplace transform or two-sided Laplace transform by extending the limits of integration to be the entire real axis. If that is done the common unilateral transform simply becomes a special case of the bilateral transform where the definition of the function being transformed is multiplied by the Heaviside step function.

The bilateral Laplace transform is defined as follows,

$$\mathcal{B}\{f\}(s) = \int_{-\infty}^{\infty} e^{-st} f(t)\, dt.$$

Inverse Laplace Transform

Two integrable functions have the same Laplace transform only if they differ on a set of Lebesgue measure zero. This means that, on the range of the transform, there is an inverse transform. In fact, besides integrable functions, the Laplace transform is a one-to-one mapping from one function space into another in many other function spaces as well, although there is usually no easy characterization of the range. Typical function spaces in which this is true include the spaces of bounded continuous functions, the space $L^\infty(0, \infty)$, or more generally tempered functions (that is, functions of at worst polynomial growth) on $(0, \infty)$. The Laplace transform is also defined and injective for suitable spaces of tempered distributions.

In these cases, the image of the Laplace transform lives in a space of analytic functions in the region of convergence. The inverse Laplace transform is given by the following complex integral, which is known by various names (the Bromwich integral, the Fourier–Mellin integral, and Mellin’s inverse formula):

$$f(t) = \mathcal{L}^{-1}\{F\}(t) = \frac{1}{2\pi i} \lim_{T\to\infty} \int_{\gamma-iT}^{\gamma+iT} e^{st} F(s)\, ds,$$

where γ is a real number so that the contour path of integration is in the region of convergence of F(s). An alternative formula for the inverse Laplace transform is given by Post's inversion formula. The limit here is interpreted in the weak-* topology.

In practice, it is typically more convenient to decompose a Laplace transform into known transforms of functions obtained from a table, and construct the inverse by inspection.

Region of Convergence

If f is a locally integrable function (or more generally a Borel measure locally of bounded variation), then the Laplace transform F(s) of f converges provided that the limit

$$\lim_{R\to\infty} \int_0^R f(t) e^{-st}\, dt$$

exists.

The Laplace transform converges absolutely if the integral

$$\int_0^\infty \left| f(t) e^{-st} \right| dt$$

exists (as a proper Lebesgue integral). The Laplace transform is usually understood as conditionally convergent, meaning that it converges in the former instead of the latter sense.

The set of values for which F(s) converges absolutely is either of the form Re(s) > a or else Re(s) ≥ a, where a is an extended real constant, $-\infty \le a \le \infty$. (This follows from the dominated convergence theorem.) The constant a is known as the abscissa of absolute convergence, and depends on the growth behavior of f(t). Analogously, the two-sided transform converges absolutely in a strip of the form a < Re(s) < b, and possibly including the lines Re(s) = a or Re(s) = b. The subset of values of s for which the Laplace transform converges absolutely is called the region of absolute convergence or the domain of absolute convergence. In the two-sided case, it is sometimes called the strip of absolute convergence. The Laplace transform is analytic in the region of absolute convergence.

Similarly, the set of values for which F(s) converges (conditionally or absolutely) is known as the region of conditional convergence, or simply the region of convergence (ROC). If the Laplace transform converges (conditionally) at $s = s_0$, then it automatically converges for all s with $Re(s) > Re(s_0)$. Therefore, the region of convergence is a half-plane of the form Re(s) > a, possibly including some points of the boundary line Re(s) = a.

In the region of convergence $Re(s) > Re(s_0)$, the Laplace transform of f can be expressed by integrating by parts as the integral

$$F(s) = (s - s_0) \int_0^\infty e^{-(s-s_0)t} \beta(t)\, dt, \quad \beta(u) = \int_0^u e^{-s_0 t} f(t)\, dt.$$

That is, in the region of convergence F(s) can effectively be expressed as the absolutely convergent Laplace transform of some other function. In particular, it is analytic.

There are several Paley–Wiener theorems concerning the relationship between the decay properties of f and the properties of the Laplace transform within the region of convergence.

In engineering applications, a function corresponding to a linear time-invariant (LTI) system is stable if every bounded input produces a bounded output. This is equivalent to the absolute convergence of the Laplace transform of the impulse response function in the region Re(s) ≥ 0. As a result, LTI systems are stable provided the poles of the Laplace transform of the impulse response function have negative real part.

This ROC is used in knowing about the causality and stability of a system.

Properties and Theorems

The Laplace transform has a number of properties that make it useful for analyzing linear dynamical systems. The most significant advantage is that differentiation and integration become multiplication and division, respectively, by s (similarly to logarithms changing multiplication of numbers to addition of their logarithms).

Because of this property, the Laplace variable s is also known as operator variable in the L domain: either derivative operator or (for s^{-1}) integration operator. The transform turns integral equations and differential equations to polynomial equations, which are much easier to solve. Once solved, use of the inverse Laplace transform reverts to the time domain.

Given the functions f(t) and g(t), and their respective Laplace transforms F(s) and G(s),

$$f(t) = \mathcal{L}^{-1}\{F(s)\},$$
$$g(t) = \mathcal{L}^{-1}\{G(s)\},$$

The following Table is a list of properties of unilateral Laplace transform:

Properties of the unilateral Laplace transform

	Time domain	s domain	Comment
Linearity	$af(t) + bg(t)$	$aF(s) + bG(s)$	Can be proved using basic rules of integration.
Frequency-domain derivative	$tf(t)$	$-F'(s)$	F' is the first derivative of F.
Frequency-domain general derivative	$t^n f(t)$	$(-1)^n F^{(n)}(s)$	More general form, nth derivative of $F(s)$.
Derivative	$f'(t)$	$sF(s) - f(0)$	f is assumed to be a differentiable function, and its derivative is assumed to be of exponential type. This can then be obtained by integration by parts
Second derivative	$f''(t)$	$s^2F(s) - sf(0) - f'(0)$	f is assumed twice differentiable and the second derivative to be of exponential type. Follows by applying the Differentiation property to $f'(t)$.
General derivative	$f^{(n)}(t)$	$s^nF(s) - \sum_{k=1}^{n} s^{n-k} f^{(k-1)}(0)$	f is assumed to be n-times differentiable, with nth derivative of exponential type. Follows by mathematical induction.
Frequency-domain integration	$\frac{1}{t}f(t)$	$\int_s^\infty F(\sigma)\,d\sigma$	This is deduced using the nature of frequency differentiation and conditional convergence.
Time-domain integration	$\int_0^t f(\tau)\,d\tau = (u * f)(t)$	$\frac{1}{s}F(s)$	$u(t)$ is the Heaviside step function. Note that $(u * f)(t)$ is the convolution of $u(t)$ and $f(t)$.
Frequency shifting	$e^{at}f(t)$	$F(s - a)$	
Time shifting	$f(t - a)u(t - a)$	$e^{-as}F(s)$	$u(t)$ is the Heaviside step function

			step function
Time scaling	$f(at)$	$\frac{1}{a}F\left(\frac{s}{a}\right)$	$a > 0$
Multiplication	$f(t)g(t)$	$\frac{1}{2\pi i}\lim_{T\to\infty}\int_{c-iT}^{c+iT} F(\sigma)G(s-\sigma)\,d\sigma$	The integration is done along the vertical line $\mathrm{Re}(\sigma) = c$ that lies entirely within the region of convergence of F.[15]
Convolution	$(f * g)(t) = \int_0^t f(\tau)g(t-\tau)\,d\tau$	$F(s)\cdot G(s)$	
Complex conjugation	$f^*(t)$	$F^*(s^*)$	
Cross-correlation	$f(t) \star g(t)$	$F^*(-s^*)\cdot G(s)$	
Periodic function	$f(t)$	$\frac{1}{1-e^{-Ts}}\int_0^T e^{-st}f(t)\,dt$	$f(t)$ is a periodic function of period T so that $f(t) = f(t+T)$, for all $t \geq 0$. This is the result of the time shifting property and the geometric series.

- Initial value theorem:

$$f(0^+) = \lim_{s\to\infty} sF(s).$$

- Final value theorem:

$f(\infty) = \lim_{s\to 0} sF(s)$, if all poles of sF(s) are in the left half-plane.

The final value theorem is useful because it gives the long-term behaviour without having to perform partial fraction decompositions or other difficult algebra. If F(s) has a pole in the right-hand plane or poles on the imaginary axis (e.g., if $f(t) = e^t$ or $f(t) = \sin(t)$), the behaviour of this formula is undefined.

Relation to Power Series

The Laplace transform can be viewed as a continuous analogue of a power series. If a(n) is a discrete function of a positive integer n, then the power series associated to a(n) is the series

$$\sum_{n=0}^{\infty} a(n)x^n$$

where x is a real variable (see Z transform). Replacing summation over n with integration over t, a continuous version of the power series becomes

$$\int_0^\infty f(t)x^t\,dt$$

where the discrete function a(n) is replaced by the continuous one f(t).

Changing the base of the power from x to e gives

$$\int_0^\infty f(t)\left(e^{\ln x}\right)^t\,dt$$

For this to converge for, say, all bounded functions f, it is necessary to require that $\ln x < 0$. Making the substitution $-s = \ln x$ gives just the Laplace transform:

$$\int_0^\infty f(t)e^{-st}\,dt$$

In other words, the Laplace transform is a continuous analog of a power series in which the discrete parameter n is replaced by the continuous parameter t, and x is replaced by e^{-s}.

Relation to Moments

The quantities

$$\mu_n = \int_0^\infty t^n f(t)\,dt$$

are the moments of the function f. If the first n moments of f converge absolutely, then by repeated differentiation under the integral,

$$(-1)^n(\mathcal{L}f)^{(n)}(0) = \mu_n.$$

This is of special significance in probability theory, where the moments of a random variable X are given by the expectation values $\mu_n = E[X^n]$. Then, the relation holds

$$\mu_n = (-1)^n \frac{d^n}{ds^n} E\left[e^{-sX}\right](0).$$

Proof of the Laplace transform of a Function's Derivative

It is often convenient to use the differentiation property of the Laplace transform to find the transform of a function's derivative. This can be derived from the basic expression for a Laplace transform as follows:

$$\begin{aligned}\mathcal{L}\{f(t)\} &= \int_{0^-}^{\infty} e^{-st} f(t)\,dt \\ &= \left[\frac{f(t)e^{-st}}{-s}\right]_{0^-}^{\infty} - \int_{0^-}^{\infty} \frac{e^{-st}}{-s} f'(t)\,dt \quad \text{(by parts)} \\ &= \left[-\frac{f(0^-)}{-s}\right] + \frac{1}{s}\mathcal{L}\{f'(t)\},\end{aligned}$$

yielding

$$\mathcal{L}\{f'(t)\} = s \cdot \mathcal{L}\{f(t)\} - f(0^-),$$

and in the bilateral case,

$$\mathcal{L}\{f'(t)\} = s \int_{-\infty}^{\infty} e^{-st} f(t)\,dt = s \cdot \mathcal{L}\{f(t)\}.$$

The general result

$$\mathcal{L}\left\{f^{(n)}(t)\right\} = s^n \cdot \mathcal{L}\{f(t)\} - s^{n-1} f(0^-) - \cdots - f^{(n-1)}(0^-),$$

where $f^{(n)}$ denotes the nth derivative of f, can then be established with an inductive argument.

Evaluating Integrals over the Positive Real Axis

A really useful property of the Laplace transform is the following:

$$\int_0^{+\infty} f(x)g(x)\,dx = \int_0^{+\infty} (\mathcal{L}f)(s) \cdot (\mathcal{L}^{-1}g)(s)\,ds$$

under suitable assumptions on the behaviour of f, g in a right neighbourhood of 0 and on the decay rate of f, g in a left neighbourhood of $+\infty$. The above formula is a variation of integration by parts, with the operators $\frac{d}{dx}$ and $\int dx$ being replaced by $\mathcal{L}$ and $\mathcal{L}^{-1}$. Let us prove the equivalent formulation:

$$\int_0^{+\infty} (\mathcal{L}f)(x)g(x)\,dx = \int_0^{+\infty} f(s)(\mathcal{L}g)(s)\,ds.$$

By plugging in $(\mathcal{L}f)(x) = \int_0^{+\infty} f(s)e^{-sx}\,ds$ the left-hand side turns into:

$$\int_0^{+\infty}\int_0^{+\infty} f(s)g(x)e^{-sx}\,ds\,dx,$$

but assuming Fubini's theorem holds, by reversing the order of integration we get the wanted right-hand side.

Evaluating Improper Integrals

Let $\mathcal{L}\{f(t)\} = F(s)$, then (see the table above)

$$\mathcal{L}\left\{\frac{f(t)}{t}\right\} = \int_s^\infty F(p)\,dp,$$

or

$$\int_0^\infty \frac{f(t)}{t}e^{-st}\,dt = \int_s^\infty F(p)\,dp.$$

Letting $s \to 0$, gives one the identity

$$\int_0^\infty \frac{f(t)}{t}\,dt = \int_0^\infty F(p)\,dp.$$

provided that the interchange of limits can be justified. Even when the interchange cannot be justified the calculation can be suggestive. For example, proceeding formally one has

$$\int_0^\infty \frac{1}{t}\left(\cos(at) - \cos(bt)\right)\,dt = \int_0^\infty \left(\frac{p}{p^2+a^2} - \frac{p}{p^2+b^2}\right)\,dp = \frac{1}{2}\ln\frac{p^2+a^2}{p^2+b^2}\bigg|_0^\infty = \ln b - \ln a.$$

The validity of this identity can be proved by other means. It is an example of a Frullani integral.

Another example is Dirichlet integral.

Relationship to Other Transforms

Laplace–Stieltjes Transform

The (unilateral) Laplace–Stieltjes transform of a function g : **R** → **R** is defined by the Lebesgue–Stieltjes integral

$$\{\mathcal{L}^* g\}(s) = \int_0^\infty e^{-st}\,dg(t).$$

The function g is assumed to be of bounded variation. If g is the antiderivative of f:

$$g(x) = \int_0^x f(t)\,dt$$

then the Laplace–Stieltjes transform of g and the Laplace transform of f coincide. In general, the Laplace–Stieltjes transform is the Laplace transform of the Stieltjes measure associated to g. So in practice, the only distinction between the two transforms is that the Laplace transform is thought of as operating on the density function of the measure, whereas the Laplace–Stieltjes transform is thought of as operating on its cumulative distribution function.

Fourier Transform

The continuous Fourier transform is equivalent to evaluating the bilateral Laplace transform with imaginary argument s = iω or s = 2πfi,

$$\begin{aligned}\hat{f}(\omega) &= \mathcal{F}\{f(t)\} \\ &= \mathcal{L}\{f(t)\}|_{s=i\omega} = F(s)|_{s=i\omega} \\ &= \int_{-\infty}^{\infty} e^{-i\omega t} f(t)\, dt\,.\end{aligned}$$

This definition of the Fourier transform requires a prefactor of 1/2 π on the reverse Fourier transform. This relationship between the Laplace and Fourier transforms is often used to determine the frequency spectrum of a signal or dynamical system.

The above relation is valid as stated if and only if the region of convergence (ROC) of F(s) contains the imaginary axis, σ = 0.

For example, the function $f(t) = \cos(\omega_0 t)$ has a Laplace transform $F(s) = s/(s^2 + \omega_0^2)$ whose ROC is Re(s) > 0. As s = iω is a pole of F(s), substituting s = iω in F(s) does not yield the Fourier transform of f(t)u(t), which is proportional to the Dirac delta-function $\delta(\omega - \omega_0)$.

However, a relation of the form

$$\lim_{\sigma \to 0^+} F(\sigma + i\omega) = \hat{f}(\omega)$$

holds under much weaker conditions. For instance, this holds for the above example provided that the limit is understood as a weak limit of measures. General conditions relating the limit of the Laplace transform of a function on the boundary to the Fourier transform take the form of Paley-Wiener theorems.

Mellin Transform

The Mellin transform and its inverse are related to the two-sided Laplace transform by a simple change of variables.

If in the Mellin transform

$$G(s) = \mathcal{M}\{g(\theta)\} = \int_0^{\infty} \theta^s g(\theta) \frac{d\theta}{\theta}$$

we set $\theta = e^{-t}$ we get a two-sided Laplace transform.

Z-Transform

The unilateral or one-sided Z-transform is simply the Laplace transform of an ideally sampled signal with the substitution of

$$z \stackrel{\text{def}}{=} e^{sT},$$

where $T = 1/f_s$ is the sampling period (in units of time e.g., seconds) and f_s is the sampling rate (in samples per second or hertz).

Let

$$\Delta_T(t) \stackrel{\text{def}}{=} \sum_{n=0}^{\infty} \delta(t - nT)$$

be a sampling impulse train (also called a Dirac comb) and

$$\begin{aligned} x_q(t) &\stackrel{\text{def}}{=} x(t)\Delta_T(t) = x(t)\sum_{n=0}^{\infty}\delta(t-nT) \\ &= \sum_{n=0}^{\infty} x(nT)\delta(t-nT) = \sum_{n=0}^{\infty} x[n]\delta(t-nT) \end{aligned}$$

be the sampled representation of the continuous-time x(t)

$$x[n] \stackrel{\text{def}}{=} x(nT) \,.$$

The Laplace transform of the sampled signal $x_{q(t)}$ is

$$\begin{aligned} X_q(s) &= \int_{0^-}^{\infty} x_q(t)e^{-st}\,dt \\ &= \int_{0^-}^{\infty} \sum_{n=0}^{\infty} x[n]\delta(t-nT)e^{-st}\,dt \\ &= \sum_{n=0}^{\infty} x[n] \int_{0^-}^{\infty} \delta(t-nT)e^{-st}\,dt \\ &= \sum_{n=0}^{\infty} x[n]e^{-nsT} \,. \end{aligned}$$

This is the precise definition of the unilateral Z-transform of the discrete function x[n]

$$X(z) = \sum_{n=0}^{\infty} x[n]z^{-n}$$

with the substitution of $z \to e^{sT}$.

Comparing the last two equations, we find the relationship between the unilateral Z-transform and the Laplace transform of the sampled signal,

$$X_q(s) = X(z)\Big|_{z=e^{sT}} \,.$$

The similarity between the Z and Laplace transforms is expanded upon in the theory of time scale calculus.

Borel Transform

The integral form of the Borel transform

$$F(s) = \int_0^\infty f(z)e^{-sz}\,dz$$

is a special case of the Laplace transform for f an entire function of exponential type, meaning that

$$|f(z)| \leq Ae^{B|z|}$$

for some constants A and B. The generalized Borel transform allows a different weighting function to be used, rather than the exponential function, to transform functions not of exponential type. Nachbin's theorem gives necessary and sufficient conditions for the Borel transform to be well defined.

Fundamental Relationships

Since an ordinary Laplace transform can be written as a special case of a two-sided transform, and since the two-sided transform can be written as the sum of two one-sided transforms, the theory of the Laplace-, Fourier-, Mellin-, and Z-transforms are at bottom the same subject. However, a different point of view and different characteristic problems are associated with each of these four major integral transforms.

Table of selected Laplace transforms

Main article: List of Laplace transforms

The following table provides Laplace transforms for many common functions of a single variable.

Because the Laplace transform is a linear operator,

- The Laplace transform of a sum is the sum of Laplace transforms of each term.

$$\mathcal{L}\{f(t) + g(t)\} = \mathcal{L}\{f(t)\} + \mathcal{L}\{g(t)\}$$

- The Laplace transform of a multiple of a function is that multiple times the Laplace transformation of that function.

$$\mathcal{L}\{af(t)\} = a\mathcal{L}\{f(t)\}$$

Using this linearity, and various trigonometric, hyperbolic, and complex number (etc.) properties and/or identities, some Laplace transforms can be obtained from others quicker than by using the definition directly.

The unilateral Laplace transform takes as input a function whose time domain is the non-negative reals, which is why all of the time domain functions in the table below are multiples of the Heaviside step function, u(t).

The entries of the table that involve a time delay τ are required to be causal (meaning that τ > 0). A causal system is a system where the impulse response h(t) is zero for all time t prior to t = 0. In general, the region of convergence for causal systems is not the same as that of anticausal systems.

Function	Time domain $f(t) = \mathcal{L}^{-1}\{F(s)\}$	Laplace s-domain $F(s) = \mathcal{L}\{f(t)\}$	Region of convergence	Reference
unit impulse	$\delta(t)$	1	all s	inspection
delayed impulse	$\delta(t-\tau)$	$e^{-\tau s}$		time shift of unit impulse
unit step	$u(t)$	$\frac{1}{s}$	$\mathrm{Re}(s) > 0$	integrate unit impulse
delayed unit step	$u(t-\tau)$	$\frac{1}{s}e^{-\tau s}$	$\mathrm{Re}(s) > 0$	time shift of unit step
ramp	$t \cdot u(t)$	$\frac{1}{s^2}$	$\mathrm{Re}(s) > 0$	integrate unit impulse twice
nth power (for integer n)	$t^n \cdot u(t)$	$\frac{n!}{s^{n+1}}$	$\mathrm{Re}(s) > 0$ $(n > -1)$	Integrate unit step n times
qth power (for complex q)	$t^q \cdot u(t)$	$\frac{\Gamma(q+1)}{s^{q+1}}$	$\mathrm{Re}(s) > 0$ $\mathrm{Re}(q) > -1$	
nth root	$\sqrt[n]{t} \cdot u(t)$	$\frac{1}{s^{\frac{1}{n}+1}}\Gamma\left(\frac{1}{n}+1\right)$	$\mathrm{Re}(s) > 0$	Set $q = 1/n$ above.
nth power with frequency shift	$t^n e^{-\alpha t} \cdot u(t)$	$\frac{n!}{(s+\alpha)^{n+1}}$	$\mathrm{Re}(s) > -\alpha$	Integrate unit step, apply frequency shift
delayed nth power with frequency shift	$(t-\tau)^n e^{-\alpha(t-\tau)} \cdot u(t-\tau)$	$\frac{n! \cdot e^{-\tau s}}{(s+\alpha)^{n+1}}$	$\mathrm{Re}(s) > -\alpha$	Integrate unit step, apply frequency shift, apply time shift
exponential decay	$e^{-\alpha t} \cdot u(t)$	$\frac{1}{s+\alpha}$	$\mathrm{Re}(s) > -\alpha$	Frequency shift of unit step
two-sided exponential decay (only for bilateral transform)	$e^{-\alpha\|t\|}$	$\frac{2\alpha}{\alpha^2 - s^2}$	$-\alpha < \mathrm{Re}(s) < \alpha$	Frequency shift of unit step
exponential approach	$(1 - e^{-\alpha t}) \cdot u(t)$	$\frac{\alpha}{s(s+\alpha)}$	$\mathrm{Re}(s) > 0$	Unit step minus exponential decay
sine	$\sin(\omega t) \cdot u(t)$	$\frac{\omega}{s^2+\omega^2}$	$\mathrm{Re}(s) > 0$	Bracewell 1978, p. 227
cosine	$\cos(\omega t) \cdot u(t)$	$\frac{s}{s^2+\omega^2}$	$\mathrm{Re}(s) > 0$	Bracewell 1978, p. 227
hyperbolic sine	$\sinh(\alpha t) \cdot u(t)$	$\frac{\alpha}{s^2-\alpha^2}$	$\mathrm{Re}(s) > \|\alpha\|$	Williams 1973, p. 88
hyperbolic cosine	$\cosh(\alpha t) \cdot u(t)$	$\frac{s}{s^2-\alpha^2}$	$\mathrm{Re}(s) > \|\alpha\|$	Williams 1973, p. 88

exponentially decaying sine wave	$e^{-\alpha t}\sin(\omega t)\cdot u(t)$	$\frac{\omega}{(s+\alpha)^2+\omega^2}$	$\mathrm{Re}(s) > -\alpha$	Bracewell 1978, p. 227
exponentially decaying cosine wave	$e^{-\alpha t}\cos(\omega t)\cdot u(t)$	$\frac{s+\alpha}{(s+\alpha)^2+\omega^2}$	$\mathrm{Re}(s) > -\alpha$	Bracewell 1978, p. 227
natural logarithm	$\ln(t)\cdot u(t)$	$-\frac{1}{s}\,[\ln(s)+\gamma]$	$\mathrm{Re}(s) > 0$	Williams 1973, p. 88
Bessel function of the first kind, of order n	$J_n(\omega t)\cdot u(t)$	$\frac{\left(\sqrt{s^2+\omega^2}-s\right)^n}{\omega^n\sqrt{s^2+\omega^2}}$	$\mathrm{Re}(s) > 0$ $(n > -1)$	Williams 1973, p. 89
Error function	$\mathrm{erf}(t)\cdot u(t)$	$\frac{1}{s}e^{\frac{1}{4}s^2}\left(1-\mathrm{erf}\frac{s}{2}\right)$	$\mathrm{Re}(s) > 0$	Williams 1973, p. 89

Explanatory notes:

- $u(t)$ represents the Heaviside step function.
- δ represents the Dirac delta function.
- $\Gamma(z)$ represents the Gamma function.
- γ is the Euler–Mascheroni constant.
- t, a real number, typically represents *time*, although it can represent *any* independent dimension.
- s is the complex frequency domain parameter, and $\mathrm{Re}(s)$ is its real part.
- α, β, τ, and ω are real numbers.
- n is an integer.

s-Domain Equivalent Circuits and Impedances

The Laplace transform is often used in circuit analysis, and simple conversions to the s-domain of circuit elements can be made. Circuit elements can be transformed into impedances, very similar to phasor impedances.

Here is a summary of equivalents:

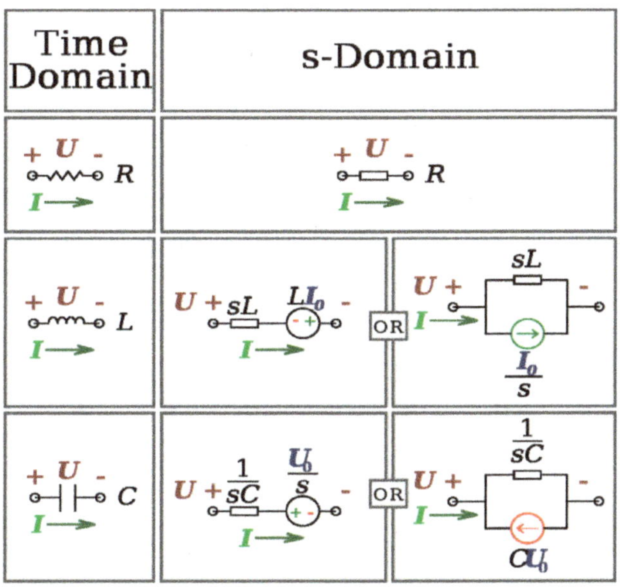

Note that the resistor is exactly the same in the time domain and the s-domain. The sources are put in if there are initial conditions on the circuit elements. For example, if a capacitor has an initial voltage across it, or if the inductor has an initial current through it, the sources inserted in the s-domain account for that.

The equivalents for current and voltage sources are simply derived from the transformations in the table above.

Examples: How to apply the properties and theorems

The Laplace transform is used frequently in engineering and physics; the output of a linear time-invariant system can be calculated by convolving its unit impulse response with the input signal. Performing this calculation in Laplace space turns the convolution into a multiplication; the latter being easier to solve because of its algebraic form.

The Laplace transform can also be used to solve differential equations and is used extensively in electrical engineering. The Laplace transform reduces a linear differential equation to an algebraic equation, which can then be solved by the formal rules of algebra. The original differential equation can then be solved by applying the inverse Laplace transform. The English electrical engineer Oliver Heaviside first proposed a similar scheme, although without using the Laplace transform; and the resulting operational calculus is credited as the Heaviside calculus.

Example 1: Solving a Differential Equation

In nuclear physics, the following fundamental relationship governs radioactive decay: the number of radioactive atoms N in a sample of a radioactive isotope decays at a rate proportional to N. This leads to the first order linear differential equation

$$\frac{dN}{dt} = -\lambda N,$$

where λ is the decay constant. The Laplace transform can be used to solve this equation.

Rearranging the equation to one side, we have

$$\frac{dN}{dt} + \lambda N = 0.$$

Next, we take the Laplace transform of both sides of the equation:

$$\left(s\tilde{N}(s) - N_0\right) + \lambda\tilde{N}(s) = 0,$$

where

$$\tilde{N}(s) = \mathcal{L}\{N(t)\}$$

and

$$N_0 = N(0).$$

Solving, we find

$$\tilde{N}(s) = \frac{N_0}{s + \lambda}.$$

Finally, we take the inverse Laplace transform to find the general solution

$$\begin{aligned} N(t) &= \mathcal{L}^{-1}\{\tilde{N}(s)\} = \mathcal{L}^{-1}\left\{\frac{N_0}{s+\lambda}\right\} \\ &= N_0 e^{-\lambda t}, \end{aligned}$$

which is indeed the correct form for radioactive decay.

Example 2: Deriving the Complex Impedance for a Capacitor

In the theory of electrical circuits, the current flow in a capacitor is proportional to the capacitance and rate of change in the electrical potential (in SI units). Symbolically, this is expressed by the differential equation

$$i = C\frac{dv}{dt},$$

where C is the capacitance (in farads) of the capacitor, i = i(t) is the electric current (in amperes) through the capacitor as a function of time, and v = v(t) is the voltage (in volts) across the terminals of the capacitor, also as a function of time.

Taking the Laplace transform of this equation, we obtain

$$I(s) = C(sV(s) - V_0),$$

where

$$\begin{aligned} I(s) &= \mathcal{L}\{i(t)\}, \\ V(s) &= \mathcal{L}\{v(t)\}, \end{aligned}$$

and

$$V_0 = v(t)|_{t=0}.$$

Solving for $V(s)$ we have

$$V(s) = \frac{I(s)}{sC} + \frac{V_0}{s}.$$

The definition of the complex impedance Z (in ohms) is the ratio of the complex voltage V divided by the complex current Iwhile holding the initial state V_0 at zero:

$$Z(s) = \left.\frac{V(s)}{I(s)}\right|_{V_0=0}.$$

Using this definition and the previous equation, we find:

$$Z(s) = \frac{1}{sC},$$

which is the correct expression for the complex impedance of a capacitor.

Example 3: Method of Partial Fraction Expansion

Consider a linear time-invariant system with transfer function

$$H(s) = \frac{1}{(s+\alpha)(s+\beta)}.$$

The impulse response is simply the inverse Laplace transform of this transfer function: $h(t) = \mathcal{L}^{-1}\{H(s)\}$.

To evaluate this inverse transform, we begin by expanding H(s) using the method of partial fraction expansion,

$$\frac{1}{(s+\alpha)(s+\beta)} = \frac{P}{s+\alpha} + \frac{R}{s+\beta}.$$

The unknown constants P and R are the residues located at the corresponding poles of the transfer function. Each residue represents the relative contribution of that singularity to the transfer function's overall shape.

By the residue theorem, the inverse Laplace transform depends only upon the poles and their residues. To find the residue P, we multiply both sides of the equation by s + α to get

$$\frac{1}{s+\beta} = P + \frac{R(s+\alpha)}{s+\beta}.$$

Then by letting s = −α, the contribution from R vanishes and all that is left is

$$P = \left.\frac{1}{s+\beta}\right|_{s=-\alpha} = \frac{1}{\beta-\alpha}.$$

Similarly, the residue R is given by

$$R = \left.\frac{1}{s+\alpha}\right|_{s=-\beta} = \frac{1}{\alpha-\beta}.$$

Note that

$$R = \frac{-1}{\beta-\alpha} = -P$$

and so the substitution of R and P into the expanded expression for H(s) gives

$$H(s) = \left(\frac{1}{\beta-\alpha}\right)\cdot\left(\frac{1}{s+\alpha} - \frac{1}{s+\beta}\right).$$

Finally, using the linearity property and the known transform for exponential decay (see Item #3 in the Table of Laplace Transforms, above), we can take the inverse Laplace transform of H(s) to obtain

$$h(t) = \mathcal{L}^{-1}\{H(s)\} = \frac{1}{\beta-\alpha}\left(e^{-\alpha t} - e^{-\beta t}\right),$$

which is the impulse response of the system.

Example 3.2: Convolution

The same result can be achieved using the convolution property as if the system is a series of filters with transfer functions of 1/(s + a) and 1/(s + b). That is, the inverse of

$$H(s) = \frac{1}{(s+a)(s+b)} = \frac{1}{s+a} \cdot \frac{1}{s+b}$$

is

$$\mathcal{L}^{-1}\left\{\frac{1}{s+a}\right\} * \mathcal{L}^{-1}\left\{\frac{1}{s+b}\right\} = e^{-at} * e^{-bt} = \int_0^t e^{-ax} e^{-b(t-x)}\, dx = \frac{e^{-at} - e^{-bt}}{b-a}.$$

Example 4: Mixing Sines, Cosines, and Exponentials

Time function	Laplace transform
$e^{-\alpha t}\left[\cos(\omega t) + \left(\frac{\beta-\alpha}{\omega}\right)\sin(\omega t)\right] u(t)$	$\frac{s+\beta}{(s+\alpha)^2+\omega^2}$

Starting with the Laplace transform

$$X(s) = \frac{s+\beta}{(s+\alpha)^2+\omega^2},$$

we find the inverse transform by first adding and subtracting the same constant α to the numerator:

$$X(s) = \frac{s+\alpha}{(s+\alpha)^2+\omega^2} + \frac{\beta-\alpha}{(s+\alpha)^2+\omega^2}.$$

By the shift-in-frequency property, we have

$$\begin{aligned} x(t) &= e^{-\alpha t}\mathcal{L}^{-1}\left\{\frac{s}{s^2+\omega^2} + \frac{\beta-\alpha}{s^2+\omega^2}\right\} \\ &= e^{-\alpha t}\mathcal{L}^{-1}\left\{\frac{s}{s^2+\omega^2} + \left(\frac{\beta-\alpha}{\omega}\right)\left(\frac{\omega}{s^2+\omega^2}\right)\right\} \\ &= e^{-\alpha t}\left[\mathcal{L}^{-1}\left\{\frac{s}{s^2+\omega^2}\right\} + \left(\frac{\beta-\alpha}{\omega}\right)\mathcal{L}^{-1}\left\{\frac{\omega}{s^2+\omega^2}\right\}\right]. \end{aligned}$$

Finally, using the Laplace transforms for sine and cosine (see the table, above), we have

$$x(t) = e^{-\alpha t}\left[\cos(\omega t)u(t) + \left(\frac{\beta-\alpha}{\omega}\right)\sin(\omega t)u(t)\right].$$

$$x(t) = e^{-\alpha t}\left[\cos(\omega t) + \left(\frac{\beta-\alpha}{\omega}\right)\sin(\omega t)\right]u(t).$$

Example 5: Phase Delay

Time function	Laplace transform
$\sin(\omega t+\phi)$	$\frac{s\sin(\phi)+\omega\cos(\phi)}{s^2+\omega^2}$
$\cos(\omega t+\phi)$	$\frac{s\cos(\phi)-\omega\sin(\phi)}{s^2+\omega^2}$.

Starting with the Laplace transform,

$$X(s) = \frac{s\sin(\phi)+\omega\cos(\phi)}{s^2+\omega^2}$$

we find the inverse by first rearranging terms in the fraction:

$$\begin{aligned} X(s) &= \frac{s\sin(\phi)}{s^2+\omega^2} + \frac{\omega\cos(\phi)}{s^2+\omega^2} \\ &= \sin(\phi)\left(\frac{s}{s^2+\omega^2}\right) + \cos(\phi)\left(\frac{\omega}{s^2+\omega^2}\right). \end{aligned}$$

We are now able to take the inverse Laplace transform of our terms:

$$\begin{aligned} x(t) &= \sin(\phi)\mathcal{L}^{-1}\left\{\frac{s}{s^2+\omega^2}\right\} + \cos(\phi)\mathcal{L}^{-1}\left\{\frac{\omega}{s^2+\omega^2}\right\} \\ &= \sin(\phi)\cos(\omega t) + \sin(\omega t)\cos(\phi). \end{aligned}$$

This is just the sine of the sum of the arguments, yielding:

$$x(t) = \sin(\omega t+\phi).$$

We can apply similar logic to find that

$$\mathcal{L}^{-1}\left\{\frac{s\cos\phi - \omega\sin\phi}{s^2+\omega^2}\right\} = \cos(\omega t+\phi).$$

Example 6: Determining Structure of Astronomical Object from Spectrum

The wide and general applicability of the Laplace transform and its inverse is illustrated by an application in astronomy which provides some information on the spatial distribution of matter of an astronomical source of radio-frequency thermal radiation too distant to resolve as more than a point, given its flux density spectrum, rather than relating the time domain with the spectrum (frequency domain).

Assuming certain properties of the object, e.g. spherical shape and constant temperature, calculations based on carrying out an inverse Laplace transformation on the spectrum of the object can produce the only possible model of the distribution of matter in it (density as a function of distance from the center) consistent with the spectrum. When independent information on the structure of an object is available, the inverse Laplace transform method has been found to be in good agreement.

FRACTIONAL CALCULUS

Fractional calculus is a branch of mathematical analysis that studies the possibility of taking real number powers, real number fractional powers or complex number powers of the differentiation operator

$$D = \frac{d}{dx},$$

and the integration operator J. (Usually J is used instead of I to avoid confusion with other I-like glyphs and identities.)

In this context, the term powers refers to iterative application of a linear operator on a function, in some analogy to function composition acting on a variable, e.g., $f^2(x) = f(f(x))$. For example, one may ask the question of meaningfully interpreting

$$\sqrt{D} = D^{\frac{1}{2}}$$

as an analog of the functional square root for the differentiation operator, i.e., an expression for some linear operator that when applied twice to any function will have the same effect as differentiation.

More generally, one can look at the question of defining the linear functional

$$D^a$$

for real-number values of a in such a way that when a takes an integer value, n, the usual power of n-fold differentiation is recovered for $n > 0$, and the −nth power of J when $n < 0$.

The motivation behind this extension to the differential operator is that the semigroup of powers D^a will form a continuous semigroup with parameter a, inside which the original discrete semigroup of D^n for integer n can be recovered as a subgroup. Continuous semigroups are prevalent in mathematics, and have an interesting theory.

Fractional differential equations (also known as extraordinary differential equations) are a generalization of differential equations through the application of fractional calculus.

Nature of the Fractional Derivative

The derivative of a function f(x) at a point x is a local property only when a is an integer; this is not the case for non-integer power derivatives. In other words, it is not correct to say that the fractional derivative at x of a function f(x) depends only on values of f very near x, in the way that integer-power derivatives certainly do. Therefore it is expected that the theory involves some sort of boundary conditions, involving information on the function further out. To use a metaphor, the fractional derivative requires some peripheral vision.

As far as the existence of such a theory is concerned, the foundations of the subject were laid by Liouville in a paper from 1832. The fractional derivative of a function to order a is often now defined by means of the Fourier or Mellin integral transforms.

Heuristics

A fairly natural question to ask is whether there exists a linear operator H, or half-derivative, such that

$$H^2 f(x) = Df(x) = \frac{d}{dx} f(x) = f'(x).$$

It turns out that there is such an operator, and indeed for any a > 0, there exists an operator P such that

$$(P^a f)(x) = f'(x),$$

or to put it another way, the definition of $d^n y/dx^n$ can be extended to all real values of n.

Let f(x) be a function defined for x > 0. Form the definite integral from 0 to x. Call this

$$(Jf)(x) = \int_0^x f(t)\, dt.$$

Repeating this process gives

$$(J^2 f)(x) = \int_0^x (Jf)(t) dt = \int_0^x \left(\int_0^t f(s)\, ds \right) dt,$$

and this can be extended arbitrarily.

The Cauchy formula for repeated integration, namely

$$(J^n f)(x) = \frac{1}{(n-1)!} \int_0^x (x-t)^{n-1} f(t)\, dt,$$

leads in a straightforward way to a generalization for real n.

Using the gamma function to remove the discrete nature of the factorial function gives us a natural candidate for fractional applications of the integral operator.

$$(J^\alpha f)(x) = \frac{1}{\Gamma(\alpha)} \int_0^x (x-t)^{\alpha-1} f(t)\, dt$$

This is in fact a well-defined operator.

It is straightforward to show that the J operator satisfies

$$(J^{\alpha})(J^{\beta}f)(x) = (J^{\beta})(J^{\alpha}f)(x) = (J^{\alpha+\beta}f)(x) = \frac{1}{\Gamma(\alpha+\beta)}\int_0^x (x-t)^{\alpha+\beta-1}f(t)\,dt$$

This relationship is called the semigroup property of fractional differintegral operators. Unfortunately the comparable process for the derivative operator D is significantly more complex, but it can be shown that D is neither commutative nor additive in general.

Fractional Derivative of a Basic Power Function

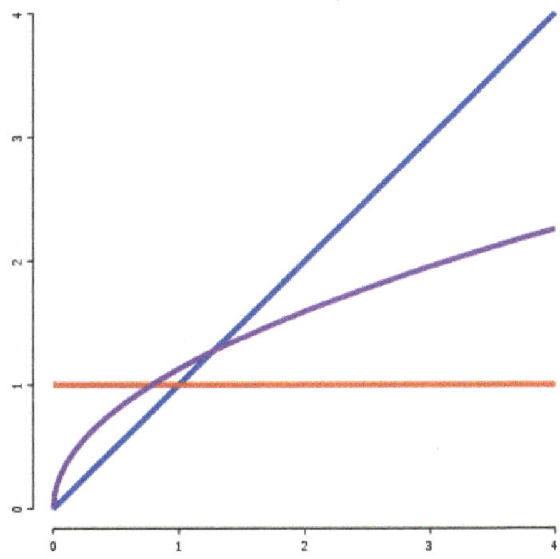

The half derivative (purple curve) of the function f(x) = x (blue curve) together with the first derivative (red curve).

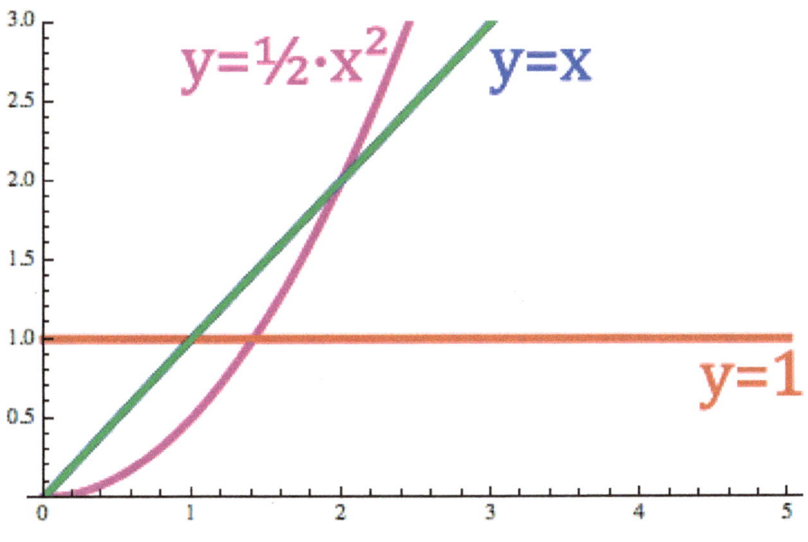

The animation shows the derivative operator oscillating between the antiderivative ($\alpha=-1$: $y = \frac{1}{2}\cdot x^2$) and the derivative ($\alpha = +1$: $y = 1$) of the simple power function $y = x$ continuously.

Let us assume that f(x) is a monomial of the form

$$f(x) = x^k .$$

The first derivative is as usual

$$f'(x) = \frac{d}{dx} f(x) = kx^{k-1} .$$

Repeating this gives the more general result that

$$\frac{d^a}{dx^a} x^k = \frac{k!}{(k-a)!} x^{k-a} ,$$

Which, after replacing the factorials with the gamma function, leads us to

$$\frac{d^a}{dx^a} x^k = \frac{\Gamma(k+1)}{\Gamma(k-a+1)} x^{k-a}, \qquad k \geq 0$$

For $k = 1$ and $a = \frac{1}{2}$, we obtain the half-derivative of the function x as

$$\frac{d^{1/2}}{dx^{1/2}} x = \frac{\Gamma(1+1)}{\Gamma(1-\frac{1}{2}+1)} x^{1-\frac{1}{2}} = \frac{\Gamma(2)}{\Gamma(\frac{3}{2})} x^{1/2} = \frac{1}{\frac{\sqrt{\pi}}{2}} x^{1/2}.$$

Repeating this process yields

$$\frac{d^{1/2}}{dx^{1/2}} \frac{2x^{\frac{1}{2}}}{\sqrt{\pi}} = \frac{2}{\sqrt{\pi}} \frac{\Gamma(1+\frac{1}{2})}{\Gamma(\frac{1}{2}-\frac{1}{2}+1)} x^{\frac{1}{2}-\frac{1}{2}} = \frac{2}{\sqrt{\pi}} \frac{\Gamma\left(\frac{3}{2}\right)}{\Gamma(1)} x^0 = \frac{2\frac{\sqrt{\pi}}{2} x^0}{\sqrt{\pi}} = 1,$$ (because $\Gamma(\frac{3}{2}) = \frac{1}{2}\sqrt{\pi}$ and $\Gamma(1) = 1$)

which is indeed the expected result of

$$\left(\frac{d^{1/2}}{dx^{1/2}} \frac{d^{1/2}}{dx^{1/2}} \right) x = \frac{d}{dx} x = 1.$$

For negative integer power k, the gamma function is undefined and we have to use the following relation:

$$\frac{d^a}{dx^a} x^{-k} = (-1)^a \frac{\Gamma(k+a)}{\Gamma(k)} x^{-(k+a)} \text{ for } k \geq 0$$

This extension of the above differential operator need not be constrained only to real powers. For example, the (1 + i)th derivative of the (1 − i)th derivative yields the 2nd derivative. Also notice that setting negative values for a yields integrals.

For a general function f(x) and 0 < α < 1, the complete fractional derivative is

$$D^\alpha f(x) = \frac{1}{\Gamma(1-\alpha)} \frac{d}{dx} \int_0^x \frac{f(t)}{(x-t)^\alpha} dt$$

For arbitrary α, since the gamma function is undefined for arguments whose real part is a negative integer and whose imaginary part is zero, it is necessary to apply the fractional derivative after the integer derivative has been performed. For example,

$$D^{3/2}f(x) = D^{1/2}D^1 f(x) = D^{1/2}\frac{d}{dx}f(x)$$

Laplace transform

We can also come at the question via the Laplace transform. Noting that

$$\mathcal{L}\left\{Jf\right\}(s) = \mathcal{L}\left\{\int_0^t f(\tau)\,d\tau\right\}(s) = \frac{1}{s}(\mathcal{L}\left\{f\right\})(s)$$

and

$$\mathcal{L}\left\{J^2 f\right\} = \frac{1}{s}(\mathcal{L}\left\{Jf\right\})(s) = \frac{1}{s^2}(\mathcal{L}\left\{f\right\})(s)$$

etc., we assert

$$J^\alpha f = \mathcal{L}^{-1}\left\{s^{-\alpha}(\mathcal{L}\{f\})(s)\right\}.$$

For example

$$J^\alpha(t^k) = \mathcal{L}^{-1}\left\{\frac{\Gamma(k+1)}{s^{\alpha+k+1}}\right\} = \frac{\Gamma(k+1)}{\Gamma(\alpha+k+1)}t^{\alpha+k}$$

as expected. Indeed, given the convolution rule

$$\mathcal{L}\{f*g\} = (\mathcal{L}\{f\})(\mathcal{L}\{g\})$$

and shorthanding $p(x) = x^{\alpha-1}$ for clarity, we find that

$$\begin{aligned}(J^\alpha f)(t) &= \frac{1}{\Gamma(\alpha)}\mathcal{L}^{-1}\left\{(\mathcal{L}\{p\})\,(\mathcal{L}\{f\})\right\}\\ &= \frac{1}{\Gamma(\alpha)}(p*f)\\ &= \frac{1}{\Gamma(\alpha)}\int_0^t p(t-\tau)f(\tau)\,d\tau\\ &= \frac{1}{\Gamma(\alpha)}\int_0^t (t-\tau)^{\alpha-1}f(\tau)\,d\tau\end{aligned}$$

which is what Cauchy gave us above.

Laplace transforms “work” on relatively few functions, but they are often useful for solving fractional differential equations.

Fractional Integrals

Atangana–Baleanu Fractional Integral

Recently, using the generalized Mittag-Leffler function, Atangana and Baleanu suggested a new formulation of the fractional derivative with nonlocal and nonsingular kernel. They then constructed the following fractional integral operator:

$$ {}_a^{AB}D_t^{-\alpha} f(t) = {}_a^{AB} I_t^{\alpha} f(t) = \frac{1-\alpha}{AB(\alpha)} f(t) + \frac{\alpha}{AB(\alpha)\Gamma(\alpha)} \int_a^t (t-\tau)^{\alpha-1} f(\tau)\, d\tau, $$

where $AB(\alpha)$ is a normalization function such that $AB(0) = AB(1) = 1$.

Riemann–Liouville Fractional Integral

The classical form of fractional calculus is given by the Riemann–Liouville integral, which is essentially what has been described above. The theory for periodic functions (therefore including the 'boundary condition' of repeating after a period) is the Weyl integral. It is defined on Fourier series, and requires the constant Fourier coefficient to vanish (thus, it applies to functions on the unit circle whose integrals evaluate to 0).

$$ {}_aD_t^{-\alpha} f(t) = {}_aI_t^{\alpha} f(t) = \frac{1}{\Gamma(\alpha)} \int_a^t (t-\tau)^{\alpha-1} f(\tau)\, d\tau $$

By contrast the Grünwald–Letnikov derivative starts with the derivative instead of the integral.

Hadamard Fractional Integral

The Hadamard fractional integral is introduced by J. Hadamard and is given by the following formula,

$$ {}_a\mathbf{D}_t^{-\alpha} f(t) = \frac{1}{\Gamma(\alpha)} \int_a^t \left(\log \frac{t}{\tau}\right)^{\alpha-1} f(\tau) \frac{d\tau}{\tau}, \qquad t > a. $$

Fractional Derivatives

Not like classical Newtonian derivatives, a fractional derivative is defined via a fractional integral.

Riemann–Liouville Fractional Derivative

The corresponding derivative is calculated using Lagrange's rule for differential operators. Computing n-th order derivative over the integral of order (n − α), the α order derivative is obtained. It is important to remark that n is the nearest integer greater than α (that is, $n = \lceil \alpha \rceil$).

$$ {}_aD_t^\alpha f(t) = \frac{d^n}{dt^n}\,{}_aD_t^{-(n-\alpha)} f(t) = \frac{d^n}{dt^n}\,{}_aI_t^{n-\alpha} f(t) $$

Caputo Fractional Derivative

There is another option for computing fractional derivatives; the Caputo fractional derivative. It was introduced by M. Caputo in his 1967 paper. In contrast to the Riemann Liouville fractional derivative, when solving differential equations using Caputo's definition, it is not necessary to define the fractional order initial conditions. Caputo's definition is illustrated as follows.

$$ {}_a^CD_t^\alpha f(t) = \frac{1}{\Gamma(n-\alpha)} \int_a^t \frac{f^{(n)}(\tau)\,d\tau}{(t-\tau)^{\alpha+1-n}}. $$

The following list summarizes the fractional derivatives defined in the literature.

Atangana–Baleanu Derivative

There is a new option of computing a fractional derivative, this option is based on the general Mittag-Leffler function as kernel. This new version was introduced by Abdon Atangana and Dumitru Baleanu in 2016 in their work. The authors introduced two versions, Atangana–Baleanu in Caputo sense (ABC) which is the convolution of a local derivative of a given function with the generalized Mittag-Leffler function. The second version is called Atangana–Baleanu fractional derivative in Riemann–Liouville sense (ABR) and is the derivative of a convolution of a given function non necessary differentiable with the generalized Mittag-Leffler function. Atangana–Baleanu fractional derivative in Caputo sense is illustrated as follows.

$$ {}_a^{ABC}D_t^\alpha f(t) = \frac{AB(\alpha)}{1-\alpha} \int_a^t f'(\tau) E_\alpha\left(-\alpha\frac{(t-\tau)^\alpha}{1-\alpha}\right). $$

And the Atangana–Baleanu fractional derivative in Riemann–Liouville is defined as:

$$ {}_a^{ABR}D_t^\alpha f(t) = \frac{AB(\alpha)}{1-\alpha} \frac{d}{dt} \int_a^t f(\tau) E_\alpha\left(-\alpha\frac{(t-\tau)^\alpha}{1-\alpha}\right). $$

Other types

Classical fractional derivatives include:

- Grünwald–Letnikov derivative
- Sonin–Letnikov derivative
- Liouville derivative
- Caputo derivative
- Hadamard derivative
- Marchaud derivative
- Riesz derivative

- Riesz–Miller derivative
- Miller–Ross derivative
- Weyl derivative
- Erdélyi–Kober derivative

New fractional derivatives include:

- Machado derivative
- Chen–Machado derivative
- Coimbra derivative
- Katugampola derivative
- Caputo–Katugampola derivative
- Hilfer derivative
- Davidson derivative
- Chen derivative
- Atangana–Baleanu derivative
- Pichaghchi derivative

Generalizations

Erdélyi–Kober Operator

The Erdélyi–Kober operator is an integral operator introduced by Arthur Erdélyi (1940). and Hermann Kober (1940) and is given by

$$\frac{x^{-\nu-\alpha+1}}{\Gamma(\alpha)}\int_0^x (t-x)^{\alpha-1}t^{-\alpha-\nu}f(t)dt,$$

which generalizes the Riemann–Liouville fractional integral and the Weyl integral.

Katugampola Operators

A recent generalization introduced by Udita Katugampola (2011) is the following, which generalizes the Riemann–Liouville fractional integral and the Hadamard fractional integral. The integral is now know as the Katugampola fractional integral and is given by,

$$\left({}^\rho\mathcal{I}^\alpha_{a+}f\right)(x)=\frac{\rho^{1-\alpha}}{\Gamma(\alpha)}\int_a^x \frac{\tau^{\rho-1}f(\tau)}{(x^\rho-\tau^\rho)^{1-\alpha}}\,d\tau, \qquad x>a.$$

Even though the integral operator in question is a close resemblance of the famous Erdélyi–Kober operator, it is not possible to obtain the Hadamard fractional integral as a direct consequence of the Erdélyi–Kober operator. Also, there is a Udita-type fractional derivative, which generalizes the Riemann–Liouville and the Hadamard fractional derivatives. As with the case of fractional integrals, the same is not true for the Erdélyi–Kober operator.

Functional Calculus

In the context of functional analysis, functions f(D) more general than powers are studied in the functional calculus of spectral theory. The theory of pseudo-differential operators also allows one to consider powers of D. The operators arising are examples of singular integral operators; and the generalization of the classical theory to higher dimensions is called the theory of Riesz potentials. So there are a number of contemporary theories available, within which fractional calculus can be discussed.

Applications

Fractional Conservation of Mass

As described by Wheatcraft and Meerschaert (2008), a fractional conservation of mass equation is needed to model fluid flow when the control volume is not large enough compared to the scale of heterogeneity and when the flux within the control volume is non-linear. In the referenced paper, the fractional conservation of mass equation for fluid flow is:

$$-\rho\left(\nabla^{\alpha}\cdot\vec{u}\right)=\Gamma(\alpha+1)\Delta x^{1-\alpha}\rho\left(\beta_s+\phi\beta_w\right)\frac{\partial p}{\partial t}$$

Groundwater Flow Problem

In 2013–2014 Atangana et al. described some groundwater flow problems using the concept of derivative with fractional order. In these works, The classical Darcy law is generalized by regarding the water flow as a function of a non-integer order derivative of the piezometric head. This generalized law and the law of conservation of mass are then used to derive a new equation for groundwater flow.

Fractional Advection Dispersion Equation

This equation has been shown useful for modeling contaminant flow in heterogenous porous media. Atangana and Kilicman extended fractional advection dispersion equation to variable order fractional advection dispersion equation. In their work, the hydrodynamic dispersion equation was generalized using the concept of variational order derivative. The modified equation was numerically solved via the Crank-Nicholson scheme. The stability and convergence of the scheme in this case were presented. The numerical simulations showed that, the modified equation is more reliable in predicting the movement of pollution in the deformable aquifers, than the constant fractional and integer derivatives

Time-Space Fractional Diffusion Equation Models

Anomalous diffusion processes in complex media can be well characterized by using fractional-order diffusion equation models. The time derivative term is corresponding to long-time heavy tail decay and the spatial derivative for diffusion nonlocality. The time-space fractional diffusion governing equation can be written as

$$\frac{\partial^\alpha u}{\partial t^\alpha} = -K(-\triangle)^\beta u.$$

A simple extension of fractional derivative is the variable-order fractional derivative, the α, β are changed into α(x, t), β(x, t). Its applications in anomalous diffusion modeling can be found in reference.

Structural Damping Models

Fractional derivatives are used to model viscoelastic damping in certain types of materials like polymers.

Acoustical Wave Equations for Complex Media

The propagation of acoustical waves in complex media, e.g. biological tissue, commonly implies attenuation obeying a frequency power-law. This kind of phenomenon may be described using a causal wave equation which incorporates fractional time derivatives:

$$\nabla^2 u - \frac{1}{c_0^2}\frac{\partial^2 u}{\partial t^2} + \tau_\sigma^\alpha \frac{\partial^\alpha}{\partial t^\alpha}\nabla^2 u - \frac{\tau_\epsilon^\beta}{c_0^2}\frac{\partial^{\beta+2} u}{\partial t^{\beta+2}} = 0.$$

Such models are linked to the commonly recognized hypothesis that multiple relaxation phenomena give rise to the attenuation measured in complex media.

Fractional Schrödinger Equation in Quantum Theory

The fractional Schrödinger equation, a fundamental equation of fractional quantum mechanics, has the following form:

$$i\hbar\frac{\partial \psi(\mathbf{r},t)}{\partial t} = \mu_\alpha(-\hbar^2\Delta)^{\frac{\alpha}{2}}\psi(\mathbf{r},t) + V(\mathbf{r},t)\psi(\mathbf{r},t).$$

where the solution of the equation is the wavefunction ψ(**r**, t) – the quantum mechanical probability amplitude for the particle to have a given position vector **r** at any given time t, and ħ is the reduced Planck constant. The potential energy function V(**r**, t) depends on the system.

Further, $\Delta = \partial^2/\partial \mathbf{r}^2$ is the Laplace operator, and μ_α is a scale constant with physical dimension $[\mu_\alpha] = J^{1-\alpha} \cdot m^\alpha \cdot s^{-\alpha} = Kg^{1-\alpha} \cdot m^{2-\alpha} \cdot s^{\alpha-2}$, (at α = 2, $\mu_2 = 1/2m$ for a particle of mass m), and the operator $(-\hbar^2\Delta)^{\alpha/2}$ is the 3-dimensional fractional quantum Riesz derivative defined by

$$-\hbar^2\Delta)^{\alpha/2}\psi(\mathbf{r},t) = \frac{1}{(2\pi\hbar)^3}\int d^3pe^{\frac{i}{\hbar}\mathbf{p}\cdot\mathbf{r}}|\mathbf{p}|^\alpha\varphi(\mathbf{p},t).$$

The index α in the fractional Schrödinger equation is the Lévy index, $1 < \alpha \leq 2$.

REFERENCES

1. Uri M. Ascher; Linda R. Petzold (1998). Computer Methods for Ordinary Differential Equations and Differential-Algebraic Equations. SIAM. p. 12. ISBN 978-1-61197-139-2.
2. Achim Ilchmann; Timo Reis (2014). Surveys in Differential-Algebraic Equations II. Springer. pp. 104–105. ISBN 978-3-319-11050-9.
3. Renate Merker and Wolfgang Schwarz, eds. (2001). System Design Automation: Fundamentals, Principles, Methods, Examples. Springer Science & Business Media. p. 221. ISBN 978-0-7923-7313-1.
4. K. E. Brenan; S. L. Campbell; L. R. Petzold (1996). Numerical Solution of Initial-value Problems in Differential-algebraic Equations. SIAM. pp. 173–177. ISBN 978-1-61197-122-4.
5. Günther, M.; Feldmann, U.; Ter Maten, J. (2005). "Modelling and Discretization of Circuit Problems". Numerical Methods in Electromagnetics. Handbook of Numerical Analysis. 13. p. 523. doi:10.1016/S1570-8659(04)13006-8. ISBN 978-0-444-51375-5., pp. 529-531
6. Ascher and Petzold, p. 234
7. Ricardo Riaza (2013). "DAEs in Circuit Modelling: A Survey". In Achim Ilchmann and Timo Reis. Surveys in Differential-Algebraic Equations I. Springer Science & Business Media. ISBN 978-3-642-34928-7.
8. Platte, D.; Jing, S.; Sommer, R.; Barke, E. (2007). "Improving Efficiency and Robustness of Analog Behavioral Models". Advances in Design and Specification Languages for Embedded Systems. p. 53. doi:10.1007/978-1-4020-6149-3_4. ISBN 978-1-4020-6147-9.
9. Hauser, M.; Salzig, C.; Dreyer, A. (2011). "Fast and Robust Symbolic Model Order Reduction with Analog Insydes". Computer Algebra in Scientific Computing. Lecture Notes in Computer Science. 6885. p. 215. doi:10.1007/978-3-642-23568-9_17. ISBN 978-3-642-23567-2.
10. Ricardo Riaza (2008). Differential-algebraic Systems: Analytical Aspects and Circuit Applications. World Scientific. pp. 5–8. ISBN 978-981-279-181-8.
11. http://www.ise.chuo-u.ac.jp/ise-labs/takamatsu-lab/takamatsu/metr/METR08-10.pdf
12. Bracewell, Ronald N. (1978), The Fourier Transform and its Applications (2nd ed.), McGraw-Hill Kogakusha, ISBN 0-07-007013-X
13. Bracewell, R. N. (2000), The Fourier Transform and Its Applications (3rd ed.),

Boston: McGraw-Hill, ISBN 0-07-116043-4

14. Feller, William (1971), An introduction to probability theory and its applications. Vol. II., Second edition, New York: John Wiley & Sons, MR 0270403
15. Korn, G. A.; Korn, T. M. (1967), Mathematical Handbook for Scientists and Engineers (2nd ed.), McGraw-Hill Companies, ISBN 0-07-035370-0
16. Widder, David Vernon (1941), The Laplace Transform, Princeton Mathematical Series, v. 6, Princeton University Press, MR 0005923
17. Williams, J. (1973), Laplace Transforms, Problem Solvers, George Allen & Unwin, ISBN 0-04-512021-8
18. Takacs, J. (1953), "Fourier amplitudok meghatarozasa operatorszamitassal", Magyar Hiradastechnika (in Hungarian), IV (7–8): 93–96
19. Euler, L. (1744), "De constructione aequationum" [The Construction of Equations], Opera omnia, 1st series (in Latin), 22: 150–161
20. Euler, L. (1753), "Methodus aequationes differentiales" [A Method for Solving Differential Equations], Opera omnia, 1st series (in Latin), 22: 181–213
21. Euler, L. (1992) [1769], "Institutiones calculi integralis, Volume 2" [Institutions of Integral Calculus], Opera omnia, 1st series (in Latin), Basel: Birkhäuser, 12, ISBN 978-3764314743, Chapters 3–5
22. Euler, Leonhard (1769), Institutiones calculi integralis [Institutions of Integral Calculus] (in Latin), II, Paris: Petropoli, ch. 3–5, pp. 57–153
23. Grattan-Guinness, I (1997), "Laplace's integral solutions to partial differential equations", in Gillispie, C. C., Pierre Simon Laplace 1749–1827: A Life in Exact Science, Princeton: Princeton University Press, ISBN 0-691-01185-0
24. Lagrange, J. L. (1773), Mémoire sur l'utilité de la méthode, Œuvres de Lagrange, 2, pp. 171–234
25. For the history of the subject, see the thesis (in French): Stéphane Dugowson, Les différentielles métaphysiques (histoire et philosophie de la généralisation de l'ordre de dérivation), Thèse, Université Paris Nord (1994)
26. Bologna, Mauro, Short Introduction to Fractional Calculus (PDF), Universidad de Tarapaca, Arica, Chile
27. Badr Saad T. Alkahtani, Chua's circuit model with Atangana–Baleanu derivative with fractional order Chaos, Solitons & Fractals, Volume 89, August 2016, Pages 547–551
28. Obaid Jefain Julaighim Algahtani. Comparing the Atangana–Baleanu and Caputo–Fabrizio derivative with fractional order: Allen Cahn modelChaos, Solitons & Fractals, Volume 89, August 2016, Pages 552–559.
29. Hadamard, J., Essai sur l'étude des fonctions données par leur développement de Taylor, Journal of pure and applied mathematics, vol. 4, no. 8, pp. 101–186, 1892.
30. Caputo, Michele (1967). Linear model of dissipation whose Q is almost frequency independent-II. Geophys. J. R. Astron. Soc. 13 (5): 529–539. doi:10.1111/j.1365-

246x.1967.tb02303.x.

31. Katugampola, U.N., A New Approach To Generalized Fractional Derivatives, Bull. Math. Anal. App. Vol 6, Issue 4, 15 October 2014, pages 1–15
32. Atangana A., Dumitru B. New fractional derivatives with non-local and non-singular kernel: Theory and application to heat transfer model, Thermal Science,Year 2016, Vol. 20, No. 2, pp. 763–769.
33. Atangana A., Koca I. Chaos in a simple nonlinear system with Atangana–Baleanu derivatives with fractional order, Chaos, Solitons & Fractals, Volume 89, August 2016, Pages 447–454. doi:10.1016/j.chaos.2016.02.012.
34. Erdélyi, Arthur (1950–51). On some functional transformations. Rendiconti del Seminario Matematico dell'Università e del Politecnico di Torino. 10: 217–234. MR 0047818.
35. Kober, Hermann (1940). On fractional integrals and derivatives. The Quarterly Journal of Mathematics (Oxford Series). 11 (1): 193–211. doi:10.1093/qmath/os-11.1.193.
36. Katugampola, U.N., New Approach To A Generalized Fractional Integral, Appl. Math. Comput. Vol 218, Issue 3, 1 October 2011, pages 860–865
37. Wheatcraft, S., Meerschaert, M., (2008). Fractional Conservation of Mass. Advances in Water Resources 31, 1377–1381.
38. Atangana, Abdon; Bildik, Necdet (2013). The Use of Fractional Order Derivative to Predict the Groundwater Flow. Mathematical Problems in Engineering. 2013: 1–9. doi:10.1155/2013/543026.
39. Atangana, Abdon; Vermeulen, P. D. (2014). Analytical Solutions of a Space-Time Fractional Derivative of Groundwater Flow Equation. Abstract and Applied Analysis. 2014: 1–11. doi:10.1155/2014/381753.
40. Benson, D., Wheatcraft, S., Meerschaert, M., (2000). Application of a fractional advection-dispersion equation. Water Resources Res 36, 1403–1412.
41. Benson, D., Wheatcraft, S., Meerschaert, M., (2000). The fractional-order governing equation of Lévy motion. Water Resources Res 36, 1413–1423.
42. Benson, D., Schumer, R., Wheatcraft, S., Meerschaert, M., (2001). Fractional dispersion, Lévy motion, and the MADE tracer tests. Transport Porous Media 42, 211–240.
43. Atangana, Abdon; Kilicman, Adem (2014). On the Generalized Mass Transport Equation to the Concept of Variable Fractional Derivative. Mathematical Problems in Engineering. 2014: 9. doi:10.1155/2014/542809.
44. Metzler, R., Klafter, J., (2000). The random walk's guide to anomalous diffusion: a fractional dynamics approach. Phys. Rep., 339, 1–77.
45. F. Mainardi, Y. Luchko, G. Pagnini, The fundamental solution of the space-time fractional diffusion equation, Fractional Calculus and Applied Analysis, Vol. 4, No 2 (2001). 153–192 arXiv:cond-mat/0702419
46. R. Gorenflo, F. Mainardi, Fractional Diffusion Processes: Probability Distributions and Continuous Time Random Walk, Springer Lecture Notes in Physics, No

621, Berlin 2003, pp. 148–166 arXiv:0709.3990

47. Fractional Calculus and Waves in Linear Viscoelasticity: An Introduction to Mathematical Models. by F. Mainardi, Imperial College Press, 2010.
48. S. Holm and S. P. Näsholm, A causal and fractional all-frequency wave equation for lossy media, Journal of the Acoustical Society of America, Volume 130, Issue 4, pp. 2195–2201 (October 2011)
49. S. P. Näsholm and S. Holm, Linking multiple relaxation, power-law attenuation, and fractional wave equations, Journal of the Acoustical Society of America, Volume 130, Issue 5, pp. 3038–3045 (November 2011).
50. S. P. Näsholm and S. Holm, On a Fractional Zener Elastic Wave Equation, Fract. Calc. Appl. Anal. Vol. 16, No 1 (2013), pp. 26–50, DOI: 10.2478/s13540-013--0003-1 Link to e-print
51. Holm S., Näsholm, S. P., Comparison of Fractional Wave Equations for Power Law Attenuation in Ultrasound and Elastography, Ultrasound Med. Biol., 40(4), pp. 695–703, DOI: 10.1016/j.ultrasmedbio.2013.09.033 Link to e-print
52. N. Laskin, (2002), Fractional Schrödinger equation, Physical Review E66, 056108 7 pages. (also available online: http://arxiv.org/abs/quant-ph/0206098)
53. Korn, G. A.; Korn, T. M. (1967), Mathematical Handbook for Scientists and Engineers (2nd ed.), McGraw-Hill Companies, ISBN 0-07-035370-0.
54. Polyanin, A. D.; Manzhirov, A. V. (1998), Handbook of Integral Equations, Boca Raton: CRC Press, ISBN 0-8493-2876-4.
55. Schwartz, Laurent (1952), "Transformation de Laplace des distributions" (in French), Comm. Sém. Math. Univ.
56. Lund [Medd. Lunds Univ. Mat. Sem.] 1952: 196–206, MR 0052555 (http:/ / www. ams. org/ mathscinet-getitem?mr=0052555).
57. Siebert, William McC. (1986), Circuits, Signals, and Systems, Cambridge, Massachusetts: MIT Press, ISBN 0-262-19229-2.
58. Widder, David Vernon (1941), The Laplace Transform, Princeton Mathematical Series, v. 6, Princeton University
59. Press, MR 0005923 (http:/ / www. ams. org/ mathscinet-getitem?mr=0005923).
60. Widder, David Vernon (1945), "What is the Laplace transform?", The American Mathematical Monthly (The American Mathematical Monthly) 52 (8): 419–425, doi: 10.2307/2305640 (http:/ / dx. doi. org/ 10. 2307/ 2305640), ISSN 0002-9890 (http:/ / www. worldcat. org/ issn/ 0002-9890), JSTOR 2305640 (http:/ / www. jstor. org/ stable/ 2305640), MR 0013447 (http:/ / www. ams. org/ mathscinet-getitem?mr=0013447).
61. Williams, J. (1973), Laplace Transforms, Problem Solvers, 10, George Allen & Unwin, ISBN 0-04-512021-8
62. Nilsson, J W, Riedel, S A (2007). Electric Circuits (8th ed.). Pearson Prentice Hall. pp. 112–113. ISBN 0-13-198925-1.

63. Ljiljana Trajković, "Nonlinear circuits", The Electrical Engineering Handbook (Ed: Wai-Kai Chen), pp. 79–81, Academic Press, 2005 ISBN 0-12-170960-4

64. US patent 3265973, Sidney Darlington, Irwin W. Sandberg, "Synthesis of two-port networks having periodically time-varying elements", issued 1966-08-09

CHAPTER 5

Linear Differential Equations

Chapter Contents

Linear differential equations are those which can be reduced to the form Ly=f, where L is some linear operator.

Your first case is indeed linear, since it can be written as:

$$\left(\frac{d^2}{dx^2} - 2\right) y = \ln(x)$$

While the second one is not. To see this first we regroup all yy to one side:

$$y(y' + 1) = x - 3$$

then we simply notice that the operator $y \mapsto g(y) = y(y' + 1)$ is not linear (for example we can take two functions y_1 and y_2 and notice that $g(y_1 + y_2) \neq g(y_1) + g(y_2)$).

INTRODUCTION

In mathematics, linear differential equations are differential equations having solutions which can be added together in particular linear combinations to form further solutions. They equate 0 to a polynomial that is linear in the value and various derivatives of a variable; its linearity means that each term in the polynomial has degree either 0 or 1.

Linear differential equations can be ordinary (ODEs) or partial (PDEs).

The solutions to (homogeneous) linear differential equations form a vector space (unlike non-linear differential equations).

BASIC FEATURES

Linear differential equations are of the form

$$Ly = f$$

where the differential operator L is a linear operator, y is the unknown function, and the right hand side f is a given function (called the source term) of the same variable. For a function dependent on time we may write the equation more expressly as

$$Ly(t) = f(t)$$

and, even more precisely by bracketing

$$L[y(t)] = f(t).$$

The linear operator L may be considered to be of the form

$$L_n(y) \equiv \frac{d^n y}{dt^n} + A_1(t)\frac{d^{n-1}y}{dt^{n-1}} + \cdots + A_{n-1}(t)\frac{dy}{dt} + A_n(t)y$$

The linearity condition on L rules out operations such as taking the square of the derivative of y; but permits, for example, taking the second derivative of y. It is convenient to rewrite this equation in an operator form

$$L_n(y) \equiv \left[D^n + A_1(t)D^{n-1} + \cdots + A_{n-1}(t)D + A_n(t)\right] y$$

where D is the differential operator d/dt (i.e. $Dy = y' = dy/dt$, $D^2y = y'' = d^2y/dt^2, \ldots$), and the A_n are given functions.

Such an equation is said to have **order** n, the index of the highest derivative of y that is involved.

A typical simple example is the linear differential equation used to model radioactive decay. Let N(t) denote the number of radioactive atoms remaining in some sample of material at time t. Then for some constant $k > 0$, the rate at which the radioactive atoms decay can be modelled by

$$\frac{dN}{dt} = -kN$$

If y is assumed to be a function of only one variable, one speaks about an ordinary differential equation, else the derivatives and their coefficients must be understood as (contracted) vectors, matrices or tensors of higher rank, and we have a (linear) partial differential equation.

The case where $f = 0$ is called a homogeneous equation and its solutions are called complementary functions. It is particularly important to the solution of the general case, since any complementary function can be added to a solution of the inhomogeneous equation to give another solution (by a method traditionally called particular integral and complementary function). When the A_i are numbers, the equation is said to have constant coefficients.

HOMOGENEOUS EQUATIONS WITH CONSTANT COEFFICIENTS

The first method of solving linear homogeneous ordinary differential equations with constant coefficients is due to Euler, who realized that solutions have the form e^{zx}, for possibly-complex values of z. The exponential function is one of the few functions to keep its shape after differentiation, allowing the sum of its multiple derivatives to cancel out to zero, as required by the equation. Thus, for constant values $A_1,..., A_n$, to solve:

$$y^{(n)} + A_1 y^{(n-1)} + \cdots + A_n y = 0\,,$$

we set $y = e^{zx}$, leading to

$$z^n e^{zx} + A_1 z^{n-1} e^{zx} + \cdots + A_n e^{zx} = 0.$$

Division by e^{zx} gives the *n*th-order polynomial:

$$F(z) = z^n + A_1 z^{n-1} + \cdots + A_n = 0.$$

This algebraic equation F(z) = 0 is the characteristic equation considered later by Gaspard Monge and Augustin-Louis Cauchy.

Formally, the terms $y^{(k)}(k = 1, 2, \ldots, n)$ of the original differential equation are replaced by z^k. Solving the polynomial gives n values of z, z_1, ..., z_n. Substitution of any of those values for z into e^{zx} gives a solution $e^{z_i x}$. Since homogeneous linear differential equations obey the superposition principle, any linear combination of these functions also satisfies the differential equation.

When these roots are all distinct, we have n distinct solutions to the differential equation. It can be shown that these are linearly independent, by applying the Vandermonde determinant, and together they form a basis of the space of all solutions of the differential equation.

Examples

$$y'''' - 2y''' + 2y'' - 2y' + y = 0$$

has the characteristic equation

$$z^4 - 2z^3 + 2z^2 - 2z + 1 = 0.$$

This has zeroes, i, $-i$, and 1 (multiplicity 2). The solution basis is then

$$e^{ix}, e^{-ix}, e^x, xe^x.$$

This corresponds to the real-valued solution basis

$$\cos x, \sin x, e^x, xe^x.$$

The preceding gave a solution for the case when all zeros are distinct, that is, each has multiplicity 1. For the general case, if z is a (possibly complex) zero (or root) of F(z) having multiplicity m, then, for $k \in \{0, 1, \ldots, m-1\}$ is a solution of the ordinary differential equation. Applying this to all roots gives a collection of n distinct and linearly independent functions, where n is the degree of F(z). As before, these functions make up a basis of the solution space.

If the coefficients A_i of the differential equation are real, then real-valued solutions are generally preferable. Since non-real roots z then come in conjugate pairs, so do their corresponding basis functions $x^k e^{zx}$, and the desired result is obtained by replacing each pair with their real-valued linear combinations Re(y) and Im(y), where y is one of the pair.

A case that involves complex roots can be solved with the aid of Euler's formula.

Second-Order Case

In the n=2 case

$$y'' + (a-1)y' + by = 0,$$

the characteristic equation is of the form

$$z^2 + (a-1)z + b = 0.$$

We then can solve for z. There are three particular cases of interest:

- Case #1: Two distinct roots, z_1 and z_2
- Case #2: One real repeated root, z
- Case #3: Complex roots, $\alpha \pm \beta i$

In case #1, the general solution is given by

$$y = c_1 e^{z_1 x} + c_2 e^{z_2 x}.$$

In case #2, the general solution is given by

$$y = (c_1 + c_2 x)e^{zx}.$$

In case #3, the general solution is given, using Euler's equation, by

$$y = c_1 e^{\alpha x} \cos(\beta x) + c_2 e^{\alpha x} \sin(\beta x),$$

$$\alpha = \mathrm{Re}(z),$$

$$\beta = \mathrm{Im}(z).$$

In each case, the constants c_1, c_2 are functions of the initial conditions $y(0)$, $y'(0)$ They can be found by using the values of the initial conditions in the solution equation for y and in the resulting equation for y' , giving two equations in the two unknown parameters.

Examples

Given $y'' - 4y' + 5y = 0$. The characteristic equation is $z^2 - 4z + 5 = 0$ which has roots "2±i". Thus the solution basis $\{y_1, y_2\}$ is $\{e^{(2+i)x}, e^{(2-i)x}\}$. Now y is a solution if and only if $y = c_1 y_1 + c_2 y_2$ for $c_1, c_2 \in \mathbf{C}$.

Because the coefficients are real,

- we are likely not interested in the complex solutions
- our basis elements are mutual conjugates

The linear combinations

$$u_1 = \mathrm{Re}(y_1) = \tfrac{1}{2}(y_1 + y_2) = e^{2x}\cos(x),$$

$$u_2 = \mathrm{Im}(y_1) = \tfrac{1}{2i}(y_1 - y_2) = e^{2x}\sin(x),$$

will give us a real basis in $\{u_1, u_2\}$.

Simple Harmonic Oscillator

The second order differential equation

$$D^2 y = -k^2 y,$$

which represents a simple harmonic oscillator, can be restated as

$$(D^2 + k^2)y = 0.$$

The expression in parenthesis can be factored out, yielding

$$((D + ik)(D - ik))y = 0,$$

which has a pair of linearly independent solutions:

$$(D - ik)y = 0$$
$$(D + ik)y = 0.$$

The solutions are, respectively,

$$y_0 = A_0 e^{ikx}$$

and

$$y_1 = A_1 e^{-ikx}.$$

These solutions provide a basis for the two-dimensional solution space of the second order differential equation: meaning that linear combinations of these solutions will also be solutions. In particular, the following solutions can be constructed

$$y_{0'} = \frac{C_0 e^{ikx} + C_0 e^{-ikx}}{2} = C_0 \cos(kx)$$

and

$$y_{1'} = \frac{C_1 e^{ikx} - C_1 e^{-ikx}}{2i} = C_1 \sin(kx).$$

These last two trigonometric solutions are linearly independent, so they can serve as another basis for the solution space, yielding the following general solution:

$$y_H = C_0 \cos(kx) + C_1 \sin(kx).$$

Damped Harmonic Oscillator

Given the equation for the damped harmonic oscillator:

$$z^2 + \frac{b}{m} z + \omega_0^2 = 0.$$

the expression in parentheses can be factored out: first obtain the characteristic equation by replacing D with z. This equation must be satisfied for all y, thus:

$$z = \tfrac{1}{2}\left(-\frac{b}{m} \pm \sqrt{\frac{b^2}{m^2} - 4\omega_0^2}\right).$$

Solve using the quadratic formula:

$$\left(D+\frac{b}{2m}-\sqrt{\frac{b^2}{4m^2}-\omega_0^2}\right)\left(D+\frac{b}{2m}+\sqrt{\frac{b^2}{4m^2}-\omega_0^2}\right)y=0.$$

This implies a pair of solutions, one corresponding to

$$\left(D+\frac{b}{2m}-\sqrt{\frac{b^2}{4m^2}-\omega_0^2}\right)y=0$$

$$\left(D+\frac{b}{2m}+\sqrt{\frac{b^2}{4m^2}-\omega_0^2}\right)y=0$$

The solutions are, respectively,

$$y_0 = A_0 e^{-\omega x+\sqrt{\omega^2-\omega_0^2}x} = A_0 e^{-\omega x} e^{\sqrt{\omega^2-\omega_0^2}x}$$

$$y_1 = A_1 e^{-\omega x-\sqrt{\omega^2-\omega_0^2}x} = A_1 e^{-\omega x} e^{-\sqrt{\omega^2-\omega_0^2}x}$$

where $\omega = b/2m$. From this linearly independent pair of solutions can be constructed another linearly independent pair which thus serve as a basis for the two-dimensional solution space:

$$y_H(A_0,A_1)(x) = \left(A_0 \sinh\left(\sqrt{\omega^2-\omega_0^2}x\right) + A_1 \cosh\left(\sqrt{\omega^2-\omega_0^2}x\right)\right)e^{-\omega x}.$$

However, if $|\omega| < |\omega_0|$ then it is preferable to get rid of the consequential imaginaries, expressing the general solution as

$$y_H(A_0,A_1)(x) = \left(A_0 \sin\left(\sqrt{\omega_0^2-\omega^2}x\right) + A_1 \cos\left(\sqrt{\omega_0^2-\omega^2}x\right)\right)e^{-\omega x}.$$

This latter solution corresponds to the underdamped case, whereas the former one corresponds to the overdamped case: the solutions for the underdamped case oscillate whereas the solutions for the overdamped case do not.

NONHOMOGENEOUS EQUATION WITH CONSTANT COEFFICIENTS

To obtain the solution to the nonhomogeneous equation (sometimes called inhomogeneous equation), find a particular integral $y_p(x)$ by either the method of undetermined coefficients or the method of variation of parameters; the general solution to the linear differential equation is the sum of the general solution of the related homogeneous equation and the particular integral. Or, when the initial conditions are set, use Laplace transform to obtain the particular solution directly.

Suppose we face

$$\frac{d^n y(x)}{dx^n} + A_1 \frac{d^{n-1} y(x)}{dx^{n-1}} + \cdots + A_n y(x) = f(x).$$

For later convenience, define the characteristic polynomial

$$P(z) = z^n + A_1 z^{n-1} + \cdots + A_n.$$

We find a solution basis $\{y_1(x), y_2(x), \ldots, y_n(x)\}$ for the homogeneous (f(x) = 0) case. We now seek a particular integral $y_p(x)$ by the variation of parameters method. Let the coefficients of the linear combination be functions of x:

$$y_p(x) = u_1(x)y_1(x) + u_2(x)y_2(x) + \cdots + u_n(x)y_n(x).$$

For ease of notation we will drop the dependency on x (i.e. the various (x)). Using the operator notation D = d/dx, the ODE in question is P(D)y = f; so

$$f = P(D)y_p = P(D)(u_1 y_1) + P(D)(u_2 y_2) + \cdots + P(D)(u_n y_n).$$

With the constraints

$$0 = u_1' y_1 + u_2' y_2 + \cdots + u_n' y_n$$
$$0 = u_1' y_1' + u_2' y_2' + \cdots + u_n' y_n'$$
$$\ldots$$
$$0 = u_1' y_1^{(n-2)} + u_2' y_2^{(n-2)} + \cdots + u_n' y_n^{(n-2)}$$

the parameters commute out,

$$f = u_1 P(D) y_1 + u_2 P(D) y_2 + \cdots + u_n P(D) y_n + u_1' y_1^{(n-1)} + u_2' y_2^{(n-1)} + \cdots + u_n' y_n^{(n-1)}.$$

But $P(D)y_j = 0$, therefore

$$f = u_1' y_1^{(n-1)} + u_2' y_2^{(n-1)} + \cdots + u_n' y_n^{(n-1)}.$$

This, with the constraints, gives a linear system in the u'_j. This much can always be solved; in fact, combining Cramer's rule with the Wronskian,

$$u_j' = (-1)^{n+j} \frac{W(y_1, \ldots, y_{j-1}, y_{j+1} \ldots, y_n)_{\binom{0}{f}}}{W(y_1, y_2, \ldots, y_n)}.$$

In the very non-standard notation used above, one should take the i,n-minor of W and multiply it by f. That's why we get a minus-sign. Alternatively, forget about the minus sign and just compute the determinant of the matrix obtained by substituting the j-th W column with (0, 0, ..., f).

The rest is a matter of integrating u'_j.

The particular integral is not unique; $y_p + c_1 y_1 + \cdots + c_n y_n$ also satisfies the ODE for any set of constants c_j.

Example

Suppose $y'' - 4y' + 5y = \sin(kx)$. We take the solution basis found above $\{e^{(2+i)x} = y_1(x), e^{(2-i)x} = y_2(x)\}$.

$$W = \begin{vmatrix} e^{(2+i)x} & e^{(2-i)x} \\ (2+i)e^{(2+i)x} & (2-i)e^{(2-i)x} \end{vmatrix} = e^{4x} \begin{vmatrix} 1 & 1 \\ 2+i & 2-i \end{vmatrix} = -2ie^{4x}$$

$$u_1' = \frac{1}{W} \begin{vmatrix} 0 & e^{(2-i)x} \\ \sin(kx) & (2-i)e^{(2-i)x} \end{vmatrix} = -\tfrac{i}{2}\sin(kx)e^{(-2-i)x}$$

$$u_2' = \frac{1}{W} \begin{vmatrix} e^{(2+i)x} & 0 \\ (2+i)e^{(2+i)x} & \sin(kx) \end{vmatrix} = \tfrac{i}{2}\sin(kx)e^{(-2+i)x}.$$

Using the list of integrals of exponential functions

$$u_1 = -\tfrac{i}{2}\int \sin(kx)e^{(-2-i)x}\,dx = \frac{ie^{(-2-i)x}}{2(3+4i+k^2)}\left((2+i)\sin(kx) + k\cos(kx)\right)$$

$$u_2 = \tfrac{i}{2}\int \sin(kx)e^{(-2+i)x}\,dx = \frac{ie^{(i-2)x}}{2(3-4i+k^2)}\left((i-2)\sin(kx) - k\cos(kx)\right).$$

And so

$$y_p = u_1(x)y_1(x) + u_2(x)y_2(x) = \frac{i}{2(3+4i+k^2)}\left((2+i)\sin(kx) + k\cos(kx)\right) + \frac{i}{2(3-4i+k^2)}\left((i-2)\sin(kx) - k\cos(kx)\right)$$
$$= \frac{(5-k^2)\sin(kx) + 4k\cos(kx)}{(3+k^2)^2 + 16}.$$

(Notice that u_1 and u_2 had factors that canceled y_1 and y_2; that is typical.)

For interest's sake, this ODE has a physical interpretation as a driven damped harmonic oscillator; y_p represents the steady state, and $c_1y_1 + c_2y_2$ is the transient.

EQUATION WITH VARIABLE COEFFICIENTS

A linear ODE of order n with variable coefficients has the general form

$$p_n(x)y^{(n)}(x) + p_{n-1}(x)y^{(n-1)}(x) + \cdots + p_0(x)y(x) = r(x).$$

Examples

A simple example is the Cauchy–Euler equation often used in engineering

$$x^n y^{(n)}(x) + a_{n-1}x^{n-1}y^{(n-1)}(x) + \cdots + a_0y(x) = 0.$$

First-order Equation with Variable Coefficients

A linear ODE of order 1 with variable coefficients has the general form

$Dy(x) + f(x)y(x) = g(x).$

Where D is the differential operator. Equations of this form can be solved by multiplying the integrating factor

$$e^{\int f(x)\,dx}$$

throughout to obtain

$$Dy(x)e^{\int f(x)\,dx} + f(x)y(x)e^{\int f(x)\,dx} = g(x)e^{\int f(x)\,dx},$$

which simplifies due to the product rule (applied backwards) to

$$D\left(y(x)e^{\int f(x)\,dx}\right) = g(x)e^{\int f(x)\,dx}$$

which, on integrating both sides and solving for $y(x)$ gives:

$$y(x) = e^{-\int f(x)\,dx}\left(\int g(x)e^{\int f(x)\,dx}\,dx + \kappa\right).$$

In other words: The solution of a first-order linear ODE

$$y'(x) + f(x)y(x) = g(x),$$

with coefficients that may or may not vary with x, is:

$$y = e^{-a(x)}\left(\int g(x)e^{a(x)}\,dx + \kappa\right)$$

where κ is the constant of integration, and

$$a(x) = \int f(x)\,dx.$$

Examples

Solve the equation

$$y'(x) + 3y(x) = 2$$

with the initial condition

$$y(0) = 2.$$

Using the general solution method:

$$y = e^{-3x}\left(\int 2e^{3x}\,dx + \kappa\right).$$

The indefinite integral is solved to give:

$$y = e^{-3x}\left((2/3)\,e^{3x} + \kappa\right).$$

Then we can reduce to:

$$y = 2/3 + \kappa e^{-3x}.$$

where $\kappa = 4/3$ from the initial condition.

A compact form of the general solution based on a Green's function is

$$y(x) = \int_a^x [y(a)\delta(t-a) + g(t)]e^{-\int_t^x f(u)du}\,dt.$$

where δ(x) is the generalized Dirac delta function.

Examples

Consider a first order differential equation with constant coefficients:

$$\frac{dy}{dx} + by = 1.$$

This equation is particularly relevant to first order systems such as RC circuits and mass-damper systems.

In this case, $f(x) = b$, $g(x) = 1$.

Hence its solution is

$$y(x) = e^{-bx}\left(\frac{e^{bx}}{b} + C\right) = \frac{1}{b} + Ce^{-bx}.$$

Systems of Linear Differential Equations

An arbitrary linear ordinary differential equation or even a system of such equations can be converted into a first order system of linear differential equations by adding variables for all but the highest order derivatives. A linear system can be viewed as a single equation with a vector-valued variable. The general treatment is analogous to the treatment above of ordinary first order linear differential equations, but with complications stemming from noncommutativity of matrix multiplication.

To solve

$$\begin{cases} \mathbf{y}'(x) & = A(x)\mathbf{y}(x) + \mathbf{b}(x) \\ \mathbf{y}(x_0) & = \mathbf{y}_0 \end{cases}$$

(here $\mathbf{y}(x)$ is a vector or matrix, and $A(x)$ is a matrix), let $U(x)$ be the solution of $\mathbf{y}'(x) = A(x)\mathbf{y}(x)$ with $U(x_0) = I$ (the identity matrix). U is a fundamental matrix for the equation — the columns of U form a complete linearly independent set of solutions for the homogeneous equation. After substituting $\mathbf{y}(x) = U(x)\mathbf{z}(x)$, the equation $\mathbf{y}'(x) = A(x)\mathbf{y}(x) + \mathbf{b}(x)$ simplifies to $U(x)\mathbf{z}'(x) = \mathbf{b}(x)$. Thus,

$$\mathbf{y}(x) = U(x)\mathbf{y_0} + U(x)\int_{x_0}^{x} U^{-1}(t)\mathbf{b}(t)\,dt$$

If $A(x_1)$ commutes with $A(x_2)$ for all x_1 and x_2, then

$$U(x) = e^{\int_{x_0}^{x} A(x)\,dx}$$

and thus

$$U^{-1}(x) = e^{-\int_{x_0}^{x} A(x)\,dx},$$

but in the general case there is no closed form solution, and an approximation method such as Magnus expansion may have to be used. Note that the exponentials are matrix exponentials.

MATRIX DIFFERENTIAL EQUATION

A differential equation is a mathematical equation for an unknown function of one or several variables that relates the values of the function itself and of its derivatives of various orders. A matrix differential equation contains more than one function stacked into vector form with a matrix relating the functions to their derivatives.

For example, a simple matrix ordinary differential equation is

$$\mathbf{x}'(t) = \mathbf{A}\mathbf{x}(t)$$

where **x**(t) is an n×1 vector of functions of an underlying variable t, **x'**(t) is the vector of first derivatives of these functions, and **A** is an matrix, of which all elements are constants.

In the case where **A** has n distinct eigenvalues, this differential equation has the following general solution,

$$\mathbf{x}(t) = c_1 e^{\lambda_1 t}\mathbf{u}_1 + c_2 e^{\lambda_2 t}\mathbf{u}_2 + \cdots + c_n e^{\lambda_n t}\mathbf{u}_n\ ,$$

where $\lambda_1, \lambda_2, ..., \lambda_n$ are the eigenvalues of **A**; $\mathbf{u}_1, \mathbf{u}_2, ..., \mathbf{u}_n$ are the respective eigenvectors of **A** ; and $c_1, c_2, ..., c_n$ are constants.

By use of the Cayley-Hamilton theorem and Vandermonde-type matrices, this formal matrix exponential solution may be reduced to a simple form. Below, this solution is displayed in terms of Putzer's algorithm.

Stability and Steady State of the Matrix System

The matrix equation

$$\mathbf{x}'(t) = \mathbf{A}\mathbf{x}(t) + \mathbf{b}$$

with n×1 parameter vector **b** is stable if and only if all eigenvalues of the matrix **A** have a negative real part. The steady state **x*** to which it converges if stable is found by setting

$$\mathbf{x}'(t) = \mathbf{0},$$

thus yielding

$$\mathbf{x}^* = -\mathbf{A}^{-1}\mathbf{b},$$

assuming **A** is invertible.

Thus, the original equation can be written in homogeneous form in terms of deviations from the steady state,

$$\mathbf{x}'(t) = \mathbf{A}[\mathbf{x}(t) - \mathbf{x}^*].$$

An equivalent way of expressing this is that **x*** is a particular solution to the non-homogeneous equation, while all solutions are in the form

$$\mathbf{x}_h + \mathbf{x}^*,$$

with $\mathbf{x}_h$ a solution to the homogeneous equation (**b**=**0**).

Stability of the Two-State-Variable Case

In the n = 2 case (with two state variables), the stability conditions that the two eigenvalues of the transition matrix A each have a negative real part are equivalent to the conditions that the trace of A be negative and its determinant be positive.

Solution in matrix form

The formal solution of $\mathbf{x}'(t) = \mathbf{A}[\mathbf{x}(t) - \mathbf{x}^*]$has the matrix exponential form

$$\mathbf{x}(t) = \mathbf{x}^* + e^{\mathbf{A}t}[\mathbf{x}(0) - \mathbf{x}^*],$$

evaluated using any of a multitude of techniques.

Putzer Algorithm for Computing e^{At}

Given a matrix $\mathbf{A}$ with eigenvalues $\lambda_1, \lambda_2, \ldots, \lambda_n$,

$$e^{\mathbf{A}t} = \sum_{j=0}^{n-1} r_{j+1}(t)\mathbf{P}_j$$

where

$$\begin{aligned}
&\mathbf{P}_0 = \mathbf{I} \\
&\mathbf{P}_j = \prod_{k=1}^{j} (\mathbf{A} - \lambda_k \mathbf{I}) = \mathbf{P}_{j-1}(\mathbf{A} - \lambda_j \mathbf{I}), \qquad j = 1, 2, \ldots, n-1 \\
&\dot{r}_1 = \lambda_1 r_1 \\
&r_1(0) = 1 \\
&\dot{r}_j = \lambda_j r_j + r_{j-1}, \qquad j = 2, 3, \ldots, n \\
&r_j(0) = 0, \qquad j = 2, 3, \ldots, n
\end{aligned}$$

The equations for $r_i(t)$ are simple first order inhomogeneous ODEs.

Note the algorithm does not require that the matrix **A** be diagonalizable and bypasses complexities of the Jordan canonical forms normally utilized.

Deconstructed example of a Matrix Ordinary Differential Equation

A first-order homogeneous matrix ordinary differential equation in two functions x(t) and y(t), when taken out of matrix form, has the following form:

$$\frac{dx}{dt} = a_1 x + b_1 y, \quad \frac{dy}{dt} = a_2 x + b_2 y$$

where a_1, a_2, b_1 and b_2 may be any arbitrary scalars.

Higher order matrix ODE's may possess a much more complicated form.

Solving Deconstructed Matrix Ordinary Differential Equations

The process of solving the above equations and finding the required functions, of this particular order and form, consists of 3 main steps. Brief descriptions of each of these steps are listed below:

- Finding the eigenvalues
- Finding the eigenvectors
- Finding the needed functions

The final, third, step in solving these sorts of ordinary differential equations is usually done by means of plugging in the values, calculated in the two previous steps into a specialized general form equation, mentioned later in this article.

Solved Example of a Matrix ODE

To solve a matrix ODE according to the three steps detailed above, using simple matrices in the process, let us find, say, a function x and a function y both in terms of the single independent variable t, in the following homogeneous linear differential equation of the first order,

$$\frac{dx}{dt} = 3x - 4y, \quad \frac{dy}{dt} = 4x - 7y\,.$$

To solve this particular ordinary differential equation system, at some point of the solution process we shall need a set of two initial values (corresponding to the two state variables at the starting point). In this case, let us pick x(0)=y(0)=1.

First Step

The first step, already mentioned above, is finding the eigenvalues of **A** in

$$\begin{pmatrix} x' \\ y' \end{pmatrix} = \begin{pmatrix} 3 & -4 \\ 4 & -7 \end{pmatrix} \begin{pmatrix} x \\ y \end{pmatrix}.$$

The derivative notation x' etc. seen in one of the vectors above is known as Lagrange's notation,(first introduced by Joseph Louis Lagrange. It is equivalent to the derivative notation dx/dt used in the previous equation, known as Leibniz's notation, honouring the name of Gottfried Leibniz.)

Once the coefficients of the two variables have been written in the matrix form **A** displayed above, one may evaluate the eigenvalues. To that end, one finds the determinant of the matrix that is formed when an identity matrix, I_n, multiplied by some constant λ, is subtracted from the above coefficient matrix to yield the characteristic polynomial of it,

$$\det\left(\begin{bmatrix} 3 & -4 \\ 4 & -7 \end{bmatrix} - \lambda \begin{bmatrix} 1 & 0 \\ 0 & 1 \end{bmatrix}\right),$$

and solve for its zeroes.

Applying further simplification and basic rules of matrix addition yields

$$\det\begin{bmatrix} 3-\lambda & -4 \\ 4 & -7-\lambda \end{bmatrix}.$$

Applying the rules of finding the determinant of a single 2×2 matrix,yields the following elementary quadratic equation,

$$\det\begin{bmatrix} 3-\lambda & -4 \\ 4 & -7-\lambda \end{bmatrix} = 0$$
$$-21 - 3\lambda + 7\lambda + \lambda^2 + 16 = 0$$

which may be reduced further to get a simpler version of the above,

$$\lambda^2 + 4\lambda - 5 = 0\,.$$

Now finding the two roots, λ_1 and λ_2 of the given quadratic equation by applying the factorization method yields

$$\lambda^2 + 5\lambda - \lambda - 5 = 0$$
$$\lambda(\lambda + 5) - 1(\lambda + 5) = 0$$
$$(\lambda - 1)(\lambda + 5) = 0$$
$$\lambda = 1, -5 .$$

The values $\lambda_1 = 1$ and $\lambda_2 = -5$, calculated above are the required eigenvalues of **A**. In some cases, say other matrix ODE's, the eigenvalues may be complex, in which case the following step of the solving process, as well as the final form and the solution, may dramatically change.

Second Step

As mentioned above, this step involves finding the eigenvectors of **A** from the information originally provided.

For each of the eigenvalues calculated we have an individual eigenvector. For the first eigenvalue, which is $\lambda_1 = 1$, we have

$$\begin{pmatrix} 3 & -4 \\ 4 & -7 \end{pmatrix}\begin{pmatrix} \alpha \\ \beta \end{pmatrix} = 1\begin{pmatrix} \alpha \\ \beta \end{pmatrix}.$$

Simplifying the above expression by applying basic matrix multiplication rules yields

$$3\alpha - 4\beta = \alpha$$
$$\alpha = 2\beta .$$

All of these calculations have been done only to obtain the last expression, which in our case is $\alpha=2\beta$. Now taking some arbitrary value, presumably a small insignificant value, which is much easier to work with, for either α or β (in most cases it does not really matter), we substitute it into $\alpha=2\beta$. Doing so produces a simple vector, which is the required eigenvector for this particular eigenvalue. In our case, we pick $\alpha=2$, which, in turn determines that $\beta=1$ and, using the standard vector notation, our vector looks like

$$\hat{\mathbf{v}}_1 = \begin{pmatrix} 2 \\ 1 \end{pmatrix}.$$

Performing the same operation using the second eigenvalue we calculated, which is $\lambda = -5$, we obtain our second eigenvector. The process of working out this vector is not shown, but the final result is

$$\hat{\mathbf{v}}_2 = \begin{pmatrix} 1 \\ 2 \end{pmatrix}.$$

Third Step

This final step actually finds the required functions that are 'hidden' behind the derivatives given to us originally. There are two functions, because our differential equations deal with two variables.

The equation, which involves all the pieces of information that we have previously found has the following form,

$$\begin{pmatrix} x \\ y \end{pmatrix} = Ae^{\lambda_1 t}\hat{\mathbf{v}}_1 + Be^{\lambda_2 t}\hat{\mathbf{v}}_2.$$

Substituting the values of eigenvalues and eigenvectors yields

$$\begin{pmatrix} x \\ y \end{pmatrix} = Ae^{t}\begin{pmatrix} 2 \\ 1 \end{pmatrix} + Be^{-5t}\begin{pmatrix} 1 \\ 2 \end{pmatrix}.$$

Applying further simplification,

$$\begin{pmatrix} 2 & 1 \\ 1 & 2 \end{pmatrix}\begin{pmatrix} Ae^{t} \\ Be^{-5t} \end{pmatrix} = \begin{pmatrix} x \\ y \end{pmatrix}.$$

Simplifying further and writing the equations for functions x and y separately,

$$x = 2Ae^{t} + Be^{-5t}$$
$$y = Ae^{t} + 2Be^{-5t}.$$

The above equations are, in fact, the general functions sought, but they are in their general form; whilst we want to actually find their exact forms and solutions, so now is the time to look back at the information given to us, the so-called initial value problem. We chose $x(0) = y(0) = 1$, which plays the role of starting point for our ordinary differential equation; application of this condition specifies the constants, A and B. As we see from the $x(0) = y(0) = 1$ condition, when t=0, the overall equation is equal to 1. Thus we may construct the following system of linear equations,

$$1 = 2A + B$$
$$1 = A + 2B.$$

Solving these equations, we find that both constants A and B equal 1/3. Therefore substituting these values into the general form of these two functions specifies their exact forms,

$$x = \frac{2}{3}e^{t} + \frac{1}{3}e^{-5t}$$
$$y = \frac{1}{3}e^{t} + \frac{2}{3}e^{-5t},$$

the two functions sought.

REFERENCES

1. Birkhoff, Garrett & Rota, Gian-Carlo (1978), Ordinary Differential Equations, New York: John Wiley and Sons, Inc., ISBN 0-471-07411-X
2. Gershenfeld, Neil (1999), The Nature of Mathematical Modeling, Cambridge, UK.: Cambridge University Press, ISBN 978-0-521-57095-4
3. Robinson, James C. (2004), An Introduction to Ordinary Differential Equations, Cambridge, UK.: Cambridge University Press, ISBN 0-521-82650-0
4. H. Moya-Cessa, F. Soto-Eguibar, DIFFERENTIAL EQUATIONS: AN OPERATIONAL APPROACH, RINTON PRESS, New Jersey, 2011. ISBN 978-1-58949-060-4
5. E. J. Putzer (1966). “Avoiding the Jordan Canonical Form in the Discussion of Linear Systems with Constant Coefficients”, The American Mathematical Monthly, 73, No. 1 (1966) 2-7. , Paper on JSTOR

CHAPTER 6

Delay Differential Equations

Chapter Contents

- **Introduction**
- **Solving DDES**
- **Reduction to ODE**
- **Functional Differential Equation**
- **Volterra Integral Equation**
- **Lotka–Volterra Equations**
- **Bifurcation Theory**

Delay differential equation or DDEs are well known to be useful in various Öelds such as age-structured population growth, control theory, and any model involving responses with non-zero delays; these include models of conveyor belts, urban tra¢ c, heat exchangers, robotics, and chatter.

The delays or lags can represent gestation times, incubation periods, transport delays, or can simply lump complicated biological processes together, accounting only for the time required for these processes to occur. Such models have the advantage of combining a simple, intuitive derivation with a wide variety of possible behavior regimes for a single system.

INTRODUCTION

In mathematics, delay differential equations (DDEs) are a type of differential equation in which the derivative of the unknown function at a certain time is given in terms of the values of the function at previous times. DDEs are also called time-delay systems, systems with aftereffect or dead-time, hereditary systems, equations with deviating argument, or differential-difference equations. They belong to the class of systems with the functional state, i.e. partial differential equations (PDEs) which are infinite dimensional, as opposed to ordinary differential equations (ODEs) having a finite dimensional state vector. Four points may give a possible explanation of the popularity of DDEs. (1) Aftereffect is an applied problem: it is well known that, together with the increasing expectations of dynamic performances, engineers need their models to behave more like the real process. Many processes include aftereffect phenomena in their inner dynamics. In addition, actuators, sensors, communication networks that are now involved in feedback control loops introduce such delays. Finally, besides actual delays, time lags are frequently used to simplify very high order models. Then, the interest for DDEs keeps on growing in all scientific areas and, especially, in control engineering. (2) Delay systems are still resistant to many classical controllers: one could think that the simplest approach would consist in replacing them by some finite-dimensional approximations. Unfortunately, ignoring effects which are adequately represented by DDEs is not a general alternative: in the best situation (constant and known delays), it leads to the same degree of complexity in the control design. In worst cases (time-varying delays, for instance), it is potentially disastrous in terms of stability and oscillations. (3) Delay properties are also surprising since several studies have shown that voluntary introduction of delays can also benefit the control. (4) In spite of their complexity, DDEs however often appear as simple infinite-dimensional models in the very complex area of partial differential equations (PDEs).

A general form of the time-delay differential equation for $x(t) \in \mathbb{R}^n$ is

$$\frac{\mathrm{d}}{\mathrm{d}t}x(t) = f(t, x(t), x_t),$$

where $x_t = \{x(\tau) : \tau \leq t\}$ represents the trajectory of the solution in the past. In this equation, f is a functional operator from $\mathbb{R} \times \mathbb{R}^n \times C^1(\mathbb{R}, \mathbb{R}^n)$ to $\mathbb{R}^n$

Examples

- Continuous delay

$$\frac{\mathrm{d}}{\mathrm{d}t}x(t) = f\left(t, x(t), \int_{-\infty}^{0} x(t+\tau)\,\mathrm{d}\mu(\tau)\right)$$

- Discrete delay

$$\frac{\mathrm{d}}{\mathrm{d}t}x(t) = f(t, x(t), x(t-\tau_1), \ldots, x(t-\tau_m)) \text{ for } \tau_1 > \cdots > \tau_m \geq 0.$$

- Linear with discrete delays

$$\frac{\mathrm{d}}{\mathrm{d}t}x(t) = A_0 x(t) + A_1 x(t-\tau_1) + \cdots + A_m x(t-\tau_m)$$

where $A_0, \ldots, A_m \in \mathbb{R}^{n \times n}$.

- Pantograph equation

$$\frac{d}{dt}x(t) = ax(t) + bx(\lambda t),$$

where a, b and λ are constants and $0 < \lambda < 1$. This equation and some more general forms are named after the pantographs on trains.

SOLVING DDES

DDEs are mostly solved in a stepwise fashion with a principle called the method of steps. For instance, consider the DDE with a single delay

$$\frac{d}{dt}x(t) = f(x(t), x(t-\tau))$$

with given initial condition $\phi: [-\tau, 0] \to \mathbb{R}^n$. Then the solution on the interval $[0, \tau]$ is given by $\psi(t)$ which is the solution to the inhomogeneous initial value problem

$$\frac{d}{dt}\psi(t) = f(\psi(t), \phi(t-\tau)),$$

with $\psi(0) = \phi(0)$ This can be continued for the successive intervals by using the solution to the previous interval as inhomogeneous term. In practice, the initial value problem is often solved numerically.

Example

Suppose $f(x(t), x(t-\tau)) = ax(t-\tau)$ and $\phi(t) = 1$. Then the initial value problem can be solved with integration,

$$x(t) = x(0) + \int_{s=0}^{t} \frac{d}{dt}x(s)\,ds = 1 + a\int_{s=0}^{t} \phi(s-\tau)\,ds,$$

i.e., $x(t) = at + 1$, where the initial condition is given by $x(0) = \phi(0) = 1$. Similarly, for the interval $t \in [\tau, 2\tau]$ we integrate and fit the initial condition,

$$\begin{aligned} x(t) = x(\tau) + \int_{s=\tau}^{t} \frac{d}{dt}x(s)\,ds &= (a\tau + 1) \\ + a\int_{s=\tau}^{t} a(s-\tau) + 1\,ds &= (a\tau+1) + a\int_{s=0}^{t-\tau} as + 1\,ds, \end{aligned}$$

i.e., $x(t) = (a\tau + 1) + a(t-\tau)\left(\frac{a(t-\tau)}{2} + 1\right)$.

REDUCTION TO ODE

In some cases, delay differential equations are equivalent to a system of ordinary differential equations.

- **Example 1** Consider an equation

$$\frac{\mathrm{d}}{\mathrm{d}t}x(t) = f\left(t, x(t), \int_{-\infty}^{0} x(t+\tau)e^{\lambda\tau}\,\mathrm{d}\tau\right).$$

Introduce $y(t) = \int_{-\infty}^{0} x(t+\tau)e^{\lambda\tau}\,\mathrm{d}\tau$ to get a system of ODEs

$$\frac{\mathrm{d}}{\mathrm{d}t}x(t) = f(t, x, y), \quad \frac{\mathrm{d}}{\mathrm{d}t}y(t) = x - \lambda y.$$

- **Example 2** An equation

$$\frac{\mathrm{d}}{\mathrm{d}t}x(t) = f\left(t, x(t), \int_{-\infty}^{0} x(t+\tau)\cos(\alpha\tau+\beta)\,\mathrm{d}\tau\right)$$

is equivalent to

$$\frac{\mathrm{d}}{\mathrm{d}t}x(t) = f(t, x, y), \quad \frac{\mathrm{d}}{\mathrm{d}t}y(t) = \cos(\beta)x + \alpha z, \quad \frac{\mathrm{d}}{\mathrm{d}t}z(t) = \sin(\beta)x - \alpha y,$$

where

$$y = \int_{-\infty}^{0} x(t+\tau)\cos(\alpha\tau+\beta)\,\mathrm{d}\tau, \quad z = \int_{-\infty}^{0} x(t+\tau)\sin(\alpha\tau+\beta)\,\mathrm{d}\tau.$$

The Characteristic Equation

Similar to ODEs, many properties of linear DDEs can be characterized and analyzed using the characteristic equation. The characteristic equation associated with the linear DDE with discrete delays

$$\frac{\mathrm{d}}{\mathrm{d}t}x(t) = A_0x(t) + A_1x(t-\tau_1) + \cdots + A_mx(t-\tau_m)$$

is

$$\det(-\lambda I + A_0 + A_1e^{-\tau_1\lambda} + \cdots + A_me^{-\tau_m\lambda}) = 0.$$

The roots λ of the characteristic equation are called characteristic roots or eigenvalues and the solution set is often referred to as the spectrum. Because of the exponential in the characteristic equation, the DDE has, unlike the ODE case, an infinite number of eigenvalues, making a spectral analysis more involved. The spectrum does however have some properties which can be exploited in the analysis. For instance, even though there are an infinite number of eigenvalues, there are only a finite number of eigenvalues to the right of any vertical line in the complex plane.

This characteristic equation is a nonlinear eigenproblem and there are many methods to compute the spectrum numerically. In some special situations it is possible to solve the characteristic equation explicitly. Consider, for example, the following DDE:

$$\frac{d}{dt}x(t) = -x(t-1).$$

The characteristic equation is

$$-\lambda - e^{-\lambda} = 0.$$

There are an infinite number of solutions to this equation for complex λ. They are given by

$$\lambda = W_k(-1),$$

where W_k is the kth branch of the Lambert W function.

Software

In MATLAB, the function dde23 can be used to numerically solve delay differential equations.

FUNCTIONAL DIFFERENTIAL EQUATION

A functional differential equation (FDE) is a differential equation with deviating argument. That is, an FDE is an equation that contains some function and some of its derivatives to different argument values.

FDEs are used in mathematical models that assume the specified behavior or phenomenon is dependent on the present as well as the past events. In other words, the past events influence the future results. For this reason, FDEs are used to in many applications rather than ordinary differential equations (ODE), which implicitly assume future behavior is independent of the past.

Definition

Unlike ODEs, which are equations containing a function of one variable and its derivative for the same input, functional differential equations are equations containing a function and some of its derivatives for different input values.

- An example of an ODE would be: $f'(x) = 2f(x) + 1$
- In comparison, an FDE would be: $f'(x) = 2f(x+3) - [f(x-1)]^2$

The simplest type of FDE, called the retarded functional differential equation (RDE) or retarded differential difference equation, is of the form

$$x'(t) = f\Big(t, x(t), x'(t-r)\Big)$$

Examples

The simplest, fundamental FDE is the linear first-order delay differential equation, also referred to as RDE, which is given by

$x'(t) = \alpha_1(t)x(t) + \alpha_2 x(t-\tau) + f(t), t \geq 0$

where α_1, α_2, τ are constants, $f(t)$ is some continuous function, and x is a scalar.

Below is a table with a comparison of ODEs and FDEs

	Ordinary differential equation	Functional differential equation
Examples	$f'(x) = x^2 - 3$	$f'(x) = 3x - f(x-4)$
	$f'(x) = f(x) - 8$	$x'(t) = 3x(2t) - \left[x(t-1)\right]^2$
	$F\left(t, x(t), x'(t), x''(t)\right) = 0$	$2x(3t+1) - 5x(4t) = 1$
	$f'(x) = 4f(x) - 3x$	

Types of FDEs

FDE is the general form for more specific types of differential equations that are used in numerous applications. There are delay differential equations, integro-differential equations, and so on.

Differential Difference Equation

Differential difference equations (DDE) are FDEs in which the argument values are discrete.

The general form for FDEs of finitely many discrete deviating arguments is shown below:

1. $x^{(n)}(t) = f\left(t, x^{(n_1)}\left(t - \tau_1(t)\right), x^{(n_2)}\left(t - \tau_2(t)\right), \ldots, x^{(n_k)}\left(t - \tau_k(t)\right)\right)$

where $x(t) \in \mathbb{R}^m$, $n_1, n_2, \ldots, n_i \geq 0$,
and $\tau_1(t), \tau_2(t), \ldots, \tau_i(t) \geq 0$

DDEs are also referred to as retarded FDEs, neutral FDEs, advanced FDEs, and mixed FDEs. This classification depends on whether the rate of change of the current state of the system depends on past values or future values or both.

Classifications of DDEs	
Retarded DDE	$x'(t) = f\left(t, x(t), x(t-\tau)\right)$
Neutral DDE	$x'(t) = f\left(t, x(t), x(t-\tau), x'(t-\tau)\right)$
Advanced DDE	$x'(t-\tau) = f\left(t, x(t), x(t-\tau)\right)$

Delay Differential Equation

Functional differential equations of retarded type are when

- $\max\{n_1, n_2, \ldots, n_k\} < n$ for the given equation (1).

In other words, this class of FDEs depends on the past and present values of the function with delays.

A simple example of an RFDE is

- $x'(t) = -x(t - \tau)$

A more general form for discrete deviating arguments can be written as

$$x'(t) = f\left(t, x\left(t - \tau_1(t)\right), x\left(t - \tau_2(t)\right), \ldots, x\left(t - \tau_k(t)\right)\right)$$

Neutral Differential Difference Equation

Functional differential equations of neutral type, or neutral differential equations (NDE) are when

- $\max\{n_1, n_2, \ldots, n_k\} < n$ for (1).

NDEs depend on past and present values of the function, similarly to RDEs, except it also depends on derivatives with delays. In other words, RDEs do not involve the given function's derivative with delays while NDEs do.

Integro-Differential Equation

Integro-differential equation (IDE) of Volterra type are FDEs in which the argument values are continuous. IDEs involve both the integrals and derivatives of some function.

The continuous integro-differential equation for RFDE $x'(t) = f\left(t, x(t - \tau_1(t)), x(t - \tau_2(t)), \ldots, x(t - \tau_k(t))\right)$ can be written as

$$x'(t) = f\left(t, \int_{t-\tau(t)}^{t} K(t, \theta, x(\theta))\right), \tau(t) \geq 0$$

Application

Functional differential equations have been used in models that determine future behavior of a certain phenomenon determined by the present and the past. Future behavior of phenomena, described by the solutions of ODEs, assumes that behavior is independent of the past. However, there can be many situations that depend on past behavior.

FDEs are applicable for models in multiple fields, such as medicine, mechanics, biology, and economics. FDEs have been used in research for heat-transfer, signal processing, evolution of a species, traffic flow and study of epidemics.

Population Growth with Time Lag

A logistic equation for population growth is given by

$$\frac{dx}{dt} = \rho x(t)\left(1 - \frac{x(t)}{k}\right),$$

where ρ is the reproduction rate and k is the carrying capacity. x(t) represents the population size at time t, and $\rho\left(1 - \frac{x(t)}{k}\right)$ is the density-dependent reproduction rate.

If we were to now apply this to an earlier time $t - \tau$, we get

$$\frac{dx}{dt} = \rho x(t)\left(1 - \frac{x(t-\tau)}{k}\right)$$

Mixing Model

Upon exposure to applications of ordinary differential equations, many come across the mixing model of some chemical solution.

Suppose there is a container holding liters of salt water. Salt water is flowing in, and out of the container at the same rate r of liters per second. In other words, the rate of water flowing in is equal to the rate of the salt water solution flowing out. Let V be the amount in liters of salt water in the container and x(t) be the uniform concentration in grams per liter of salt water at time t. Then, we have the differential equation

$$x'(t) = -\frac{r}{V}x(t), \frac{r}{V} > 0$$

The problem with this equation is that it makes the assumption that every drop of water that enters the contain is instantaneously mixed into the solution. This can be eliminated by using a FDE instead of an ODE.

Let x(t) be the average concentration at time t, rather than uniform. Then, let's assume the solution leaving the container at time t is equal to $x(t-\tau), \tau > 0$, the average concentration at some earlier time. Then, the equation is a delay-differential equation of the form

$$x'(t) = -\frac{r}{V}x(t-\tau)$$

Volterra's Predator-Prey Model

The Lotka–Volterra predator-prey model was originally developed to observe the population of sharks and fish in the Adriatic Sea; however, this model has been used in many other fields for different uses, such as describing chemical reactions. Modelling predatory-prey population has always been widely researched, and as a result, there have been many different forms of the original equation.

One example, as shown by Xu, Wu (2013), of the Lotka–Volterra model with time-delay is given below:

$$p'(t) = p(t)\left[r_1(t) - a_{11}(t)p\Big(t - \tau_{11}(t)\Big) - a_{12}(t)P_1\Big(t - \tau_{12}(t)\Big) - a_{13}(t)P_2\Big(t - \tau_{13}(t)\Big)\right]$$

$$P_1'(t) = P_1(t)\left[-r_2(t) + a_{21}(t)p\Big(t - \tau_{21}(t)\Big) - a_{22}(t)P_1\Big(t - \tau_{22}(t)\Big) - a_{23}(t)P_2\Big(t - \tau_{23}(t)\Big)\right]$$

$$P_2'(t) = P_2(t)\left[-r_2(t) + a_{31}(t)p\Big(t - \tau_{31}(t)\Big) - a_{32}(t)P_1\Big(t - \tau_{32}(t)\Big) - a_{33}(t)P_2\Big(t - \tau_{33}(t)\Big)\right]$$

where $p(t)$ denotes the prey population density at time t, $P_1(t)$ and $P_2(t)$ denote the density of the predator population at time $t, r_i, a_{ij} \in C(\mathbb{R}, [0, \infty)$ and $\tau_{ij} \in C(\mathbb{R}, \mathbb{R})$

Note that this model uses linear partial differential equations.

Other Models using FDEs

Examples of other models that have used FDEs, namely RFDEs, are given below:

- Controlled motion of a rigid body
- Periodic motions
- Flip-flop circuit as a NDE
- Model of HIV epidemic
- Math models of sugar quantity in blood
- Evolution equations of single species
- Spread of an infection between two species

VOLTERRA INTEGRAL EQUATION

In mathematics, the Volterra integral equations are a special type of integral equations. They are divided into two groups referred to as the first and the second kind.

A linear Volterra equation of the first kind is

$$f(t) = \int_a^t K(t,s)\,x(s)\,ds$$

where f is a given function and x is an unknown function to be solved for. A linear Volterra equation of the second kind is

$$x(t) = f(t) + \int_a^t K(t,s)x(s)\,ds.$$

In operator theory, and in Fredholm theory, the corresponding operators are called Volterra operators.

A linear Volterra integral equation is a convolution equation if

$$x(t) = f(t) + \int_{t_0}^{t} K(t-s)x(s)\,ds.$$

The function K in the integral is called the kernel. Such equations can be analysed and solved by means of Laplace transform techniques.

The Volterra integral equations were introduced by Vito Volterra and then studied by Traian Lalescu in his 1908 thesis, Sur les équations de Volterra, written under the direction of Émile Picard. In 1911, Lalescu wrote the first book ever on integral equations.

Volterra integral equations find application in demography, the study of viscoelastic materials, and in insurance mathematics through the renewal equation.

LOTKA–VOLTERRA EQUATIONS

The Lotka–Volterra equations, also known as the predator–prey equations, are a pair of first-order, non-linear, differential equations frequently used to describe the dynamics of biological systems in which two species interact, one as a predator and the other as prey. The populations change through time according to the pair of equations:

$$\frac{dx}{dt} = \alpha x - \beta xy$$

$$\frac{dy}{dt} = \delta xy - \gamma y$$

where

- x is the number of prey (for example, rabbits);
- y is the number of some predator (for example, foxes);
- $\frac{dy}{dt}$ and $\frac{dx}{dt}$ represent the growth rates of the two populations over time;
- t represents time; and
- α, β, γ, δ are positive real parameters describing the interaction of the two species.

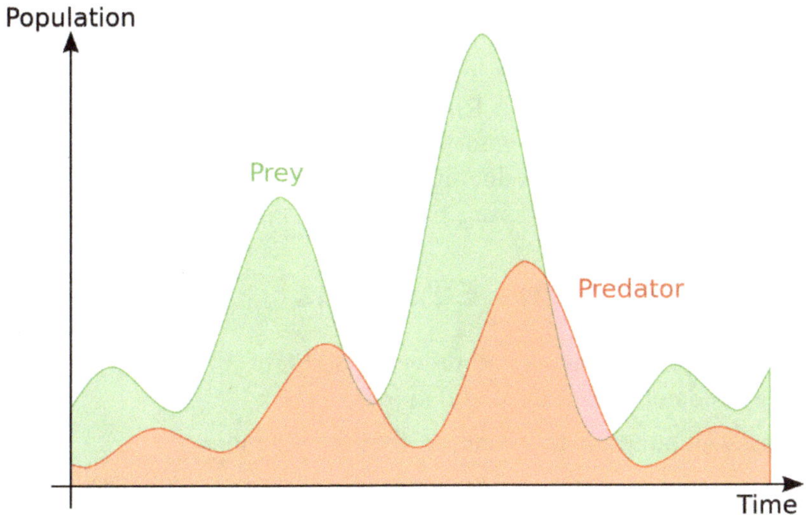

The Lotka–Volterra system of equations is an example of a Kolmogorov model, which is a more general framework that can model the dynamics of ecological systems with predator–prey interactions, competition, disease, and mutualism.

History

The Lotka–Volterra predator–prey model was initially proposed by Alfred J. Lotka in the theory of autocatalytic chemical reactions in 1910. This was effectively the logistic equation,which was originally derived by Pierre François Verhulst. In 1920 Lotka extended, via Kolmogorov (see above), the model to "organic systems" using a plant species and a herbivorous animal species as an example and in 1925 he utilised the equations to analyse predator-prey interactions in his book on biomathematics. The same set of equations were published in 1926 by Vito Volterra, a mathematician and physicist who had become interested in mathematical biology. Volterra's enquiry was inspired through his interactions with the marine biologist Umberto D'Ancona who was courting his daughter at the time and later was to become his son-in-law. D'Ancona studied the fish catches in the Adriatic Sea and had noticed that the percentage of predatory fish caught had increased during the years of World War I (1914–18). This puzzled him as the fishing effort had been very much reduced during the war years. Volterra developed his model independently from Lotka and used it to explain d'Ancona's observation.

The model was later extended to include density dependent prey growth and a functional response of the form developed by C.S. Holling; a model that has become known as the Rosenzweig-McArthur model. Both the Lotka–Volterra and Rosenzweig-MacArthur models have been used to explain the dynamics of natural populations of predators and prey, such as the lynx and snowshoe hare data of the Hudson Bay Company and the moose and wolf populations in Isle Royale National Park.

In the late 1980s an alternative to the Lotka–Volterra predator-prey model (and its common prey dependent generalizations) emerged, the ratio dependent or Arditi–Ginzburg model.The validity of prey or ratio dependent models has been much debated.

In Economics

The Lotka–Volterra equations have a long history of use in economic theory; their initial application is commonly credited to Richard Goodwin in 1965 or 1967. In economics, links are between many if not all industries; a proposed way to model the dynamics of various industries has been by introducing trophic functions between various sectors, and ignoring smaller sectors by considering the interactions of only two industrial sectors.

Physical Meaning of the Equations

The Lotka–Volterra model makes a number of assumptions about the environment and evolution of the predator and prey populations:

- The prey population finds ample food at all times.
- The food supply of the predator population depends entirely on the size of the prey population.
- The rate of change of population is proportional to its size.
- During the process, the environment does not change in favour of one species and genetic adaptation is inconsequential.
- Predators have limitless appetite.

As differential equations are used, the solution is deterministic and continuous. This, in turn, implies that the generations of both the predator and prey are continually overlapping.

Prey

When multiplied out, the prey equation becomes

$$\frac{dx}{dt} = \alpha x - \beta xy\,.$$

The prey are assumed to have an unlimited food supply, and to reproduce exponentially unless subject to predation; this exponential growth is represented in the equation above by the term αx. The rate of predation upon the prey is assumed to be proportional to the rate at which the predators and the prey meet; this is represented above by βxy. If either x or y is zero then there can be no predation.

With these two terms the equation above can be interpreted as: the change in the prey's numbers is given by its own growth minus the rate at which it is preyed upon.

Predators

The predator equation becomes

$$\frac{dy}{dt} = \delta xy - \gamma y \, .$$

In this equation, δxy represents the growth of the predator population. (Note the similarity to the predation rate; however, a different constant is used as the rate at which the predator population grows is not necessarily equal to the rate at which it consumes the prey). γy represents the loss rate of the predators due to either natural death or emigration; it leads to an exponential decay in the absence of prey.

Hence the equation expresses the change in the predator population as growth fueled by the food supply, minus natural death.

Solutions to the Equations

The equations have periodic solutions and do not have a simple expression in terms of the usual trigonometric functions, although they are quite tractable.

If none of the non-negative parameters α,β,γ,δ vanishes, three can be absorbed into the normalization of variables to leave but merely one behind: Since the first equation is homogeneous in x, and the second one in y, the parameters β/α and δ/γ, are absorbable in the normalizations of y and x, respectively, and γ into the normalization of t, so that only α/γremains arbitrary. It is the only parameter affecting the nature of the solutions.

A linearization of the equations yields a solution similar to simple harmonic motion with the population of predators trailing that of prey by 90° in the cycle.

Frequency Plot

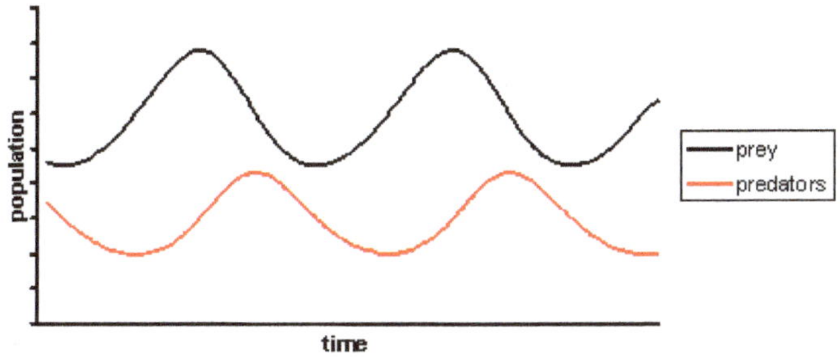

A Simple Example

Suppose there are two species of animals, a baboon (prey) and a cheetah (predator). If the initial conditions are 80 baboons and 40 cheetahs, one can plot the progression of the two species over time. The choice of time interval is arbitrary.

Frequency Plot

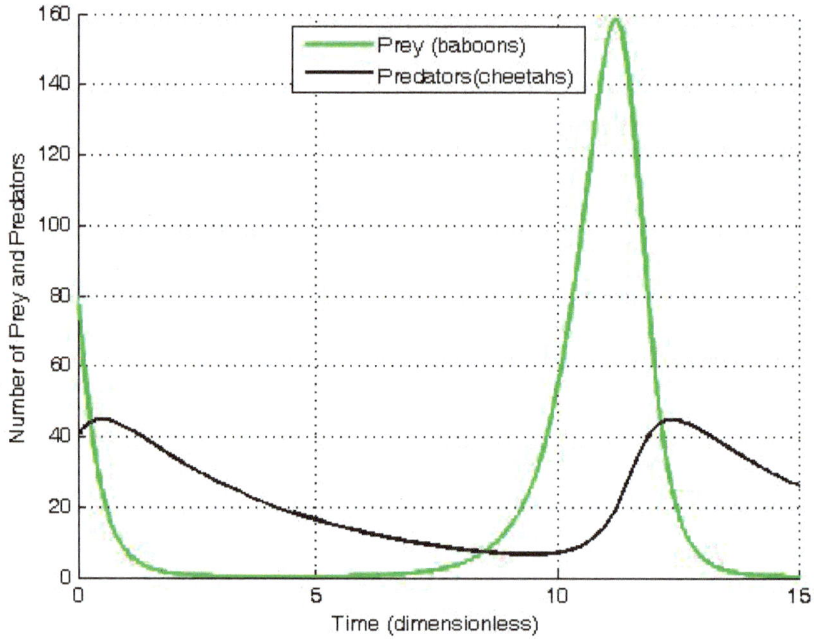

One may also plot solutions parametrically as orbits in "phase-space", without representing time, but with one axis representing the number of prey and the other axis representing the number of predators for all times.

This is to say, eliminating time from the two differential equations above results in only one such,

$$\frac{dy}{dx} = -\frac{y}{x}\frac{\delta x - \gamma}{\beta y - \alpha},$$

whose solutions are closed curves. It is amenable to separation of variables: integrating

$$d\ln y \ (\alpha - \beta y) - d\ln x \ (\gamma - \delta x) = 0$$

yields an evident constant quantity V depending on the initial conditions, which is conserved on each curve,

$$V = -\delta\, x + \gamma \ln(x) - \beta\, y + \alpha \ln(y)\,.$$

Phase Space Plot

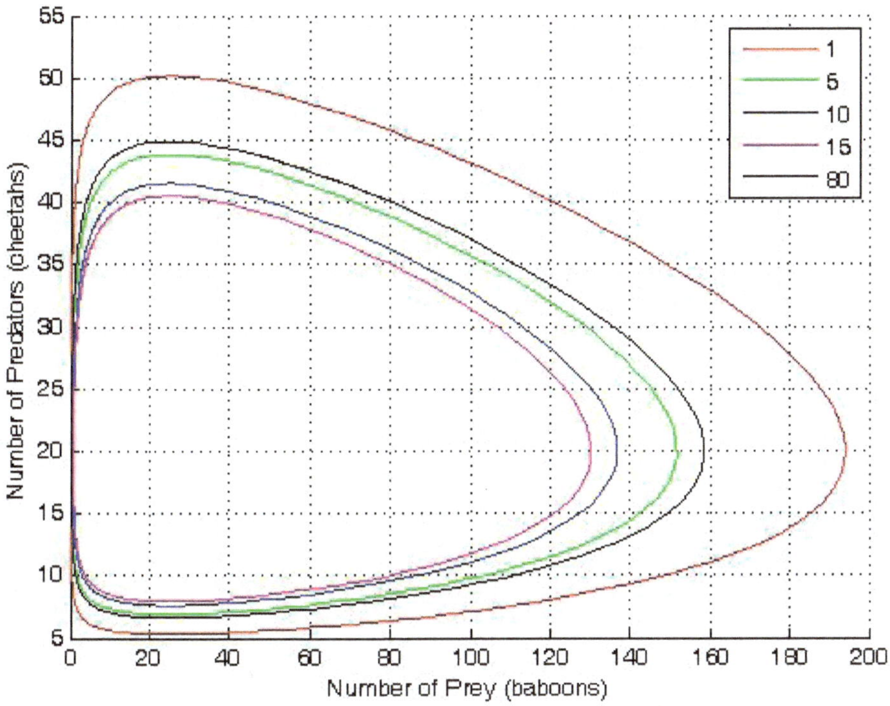

An aside: These graphs illustrate a serious potential problem with this as a biological model: For this specific choice of parameters, in each cycle, the baboon population is reduced to extremely low numbers, yet recovers (while the cheetah population remains sizeable at the lowest baboon density). In real-life situations, however, chance fluctuations of the discrete numbers of individuals, as well as the family structure and life-cycle of baboons, might cause the baboons to actually go extinct, and, by consequence, the cheetahs as well. This modelling problem has been called the "atto-fox problem", an atto-fox being a notional 10^{-18} of a fox, in the context of rabies modelling in the UK.

Phase-Space Plot of a Further Example

A less extreme example covers: $\alpha = 2/3$, $\beta = 4/3$, $\gamma = 1 = \delta$. Assume x,y quantify thousands, each. Circles represent prey and predator initial conditions from x = y =0.9 to 1.8, in steps of 0.1. The fixed point is at (1,1/2).

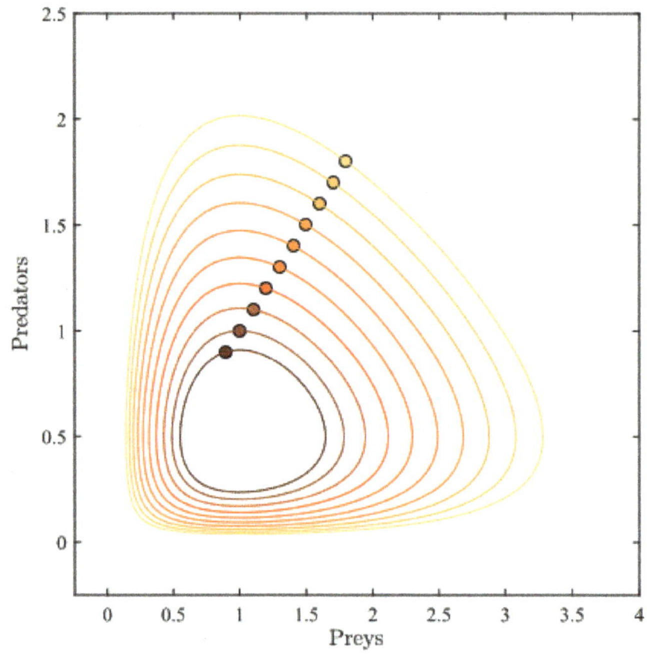

Dynamics of the system

In the model system, the predators thrive when there are plentiful prey but, ultimately, outstrip their food supply and decline. As the predator population is low the prey population will increase again. These dynamics continue in a cycle of growth and decline.

Population Equilibrium

Population equilibrium occurs in the model when neither of the population levels is changing, i.e. when both of the derivatives are equal to 0.

$$x(\alpha - \beta y) = 0$$

$$-y(\gamma - \delta x) = 0$$

When solved for x and y the above system of equations yields

$$\{y = 0, x = 0\}$$

and

$$\left\{y = \frac{\alpha}{\beta}, x = \frac{\gamma}{\delta}\right\},$$

Hence, there are two equilibria.

The first solution effectively represents the extinction of both species. If both populations are at 0, then they will continue to be so indefinitely. The second solution represents a fixed point at which both populations sustain their current, non-zero numbers, and, in the simplified model, do so indefinitely. The levels of population at which this equilibrium is achieved depend on the chosen values of the parameters, α, β, γ, and δ.

Stability of the Fixed Points

The stability of the fixed point at the origin can be determined by performing a linearization using partial derivatives.

The Jacobian matrix of the predator-prey model is

$$J(x,y) = \begin{bmatrix} \alpha - \beta y & -\beta x \\ \delta y & \delta x - \gamma \end{bmatrix}.$$

First Fixed Point (Extinction)

When evaluated at the steady state of (0, 0) the Jacobian matrix J becomes

$$J(0,0) = \begin{bmatrix} \alpha & 0 \\ 0 & -\gamma \end{bmatrix}.$$

The eigenvalues of this matrix are

$$\lambda_1 = \alpha, \quad \lambda_2 = -\gamma.$$

In the model α and γ are always greater than zero, and as such the sign of the eigenvalues above will always differ. Hence the fixed point at the origin is a saddle point.

The stability of this fixed point is of significance. If it were stable, non-zero populations might be attracted towards it, and as such the dynamics of the system might lead towards the extinction of both species for many cases of initial population levels. However, as the fixed point at the origin is a saddle point, and hence unstable, it follows that the extinction of both species is difficult in the model. (In fact, this could only occur if the prey were artificially completely eradicated, causing the predators to die of starvation. If the predators were eradicated, the prey population would grow without bound in this simple model): The populations of prey and predator can get infinitesimally close to zero and still recover.

Second Fixed Point (Oscillations)

Evaluating J at the second fixed point leads to

$$J\left(\frac{\gamma}{\delta}, \frac{\alpha}{\beta}\right) = \begin{bmatrix} 0 & -\frac{\beta\gamma}{\delta} \\ \frac{\alpha\delta}{\beta} & 0 \end{bmatrix}.$$

The eigenvalues of this matrix are

$$\lambda_1 = i\sqrt{\alpha\gamma}, \quad \lambda_2 = -i\sqrt{\alpha\gamma}.$$

As the eigenvalues are both purely imaginary and conjugate to each others, this fixed point is elliptic, so the solutions are periodic, oscillating on a small ellipse around the fixed point, with a period $\omega = \sqrt{\lambda_1 \lambda_1} = \sqrt{\alpha\gamma}$.

As illustrated in the circulating oscillations in the figure above, the level curves are closed orbits surrounding the fixed point: the levels of the predator and prey populations cycle, and oscillate without damping around the fixed point with period $\omega = \sqrt{\alpha\gamma}$.

The value of the constant of motion V, or, equivalently, K=exp(V), $:K = y^\alpha e^{-\beta y} x^\gamma e^{-\delta x}$ can be found for the closed orbits near the fixed point.

Increasing K moves a closed orbit closer to the fixed point. The largest value of the constant K is obtained by solving the optimization problem

$$y^\alpha e^{-\beta y} x^\gamma e^{-\delta x} = \frac{y^\alpha x^\gamma}{e^{\delta x + \beta y}} \longrightarrow \max_{x,y>0}.$$

The maximal value of K is thus attained at the stationary (fixed) point $\left(\frac{\gamma}{\delta}, \frac{\alpha}{\beta}\right)$ and amounts to

$$K^* = \left(\frac{\alpha}{\beta e}\right)^\alpha \left(\frac{\gamma}{\delta e}\right)^\gamma,$$

where e is Euler's Number.

BIFURCATION THEORY

Bifurcation theory is the mathematical study of changes in the qualitative or topological structure of a given family, such as the integral curves of a family of vector fields, and the solutions of a family of differential equations. Most commonly applied to the mathematical study of dynamical systems, a bifurcation occurs when a small smooth change made to the parameter values (the bifurcation parameters) of a system causes a sudden ‹qualitative' or topological change in its behavior. Bifurcations occur in both continuous systems (described by ODEs, DDEs or PDEs) and discrete systems (described by maps). The name "bifurcation" was first introduced by Henri Poincaré in 1885 in the first paper in mathematics showing such a behavior. Henri Poincaré also later named various types of stationary points and classified them.

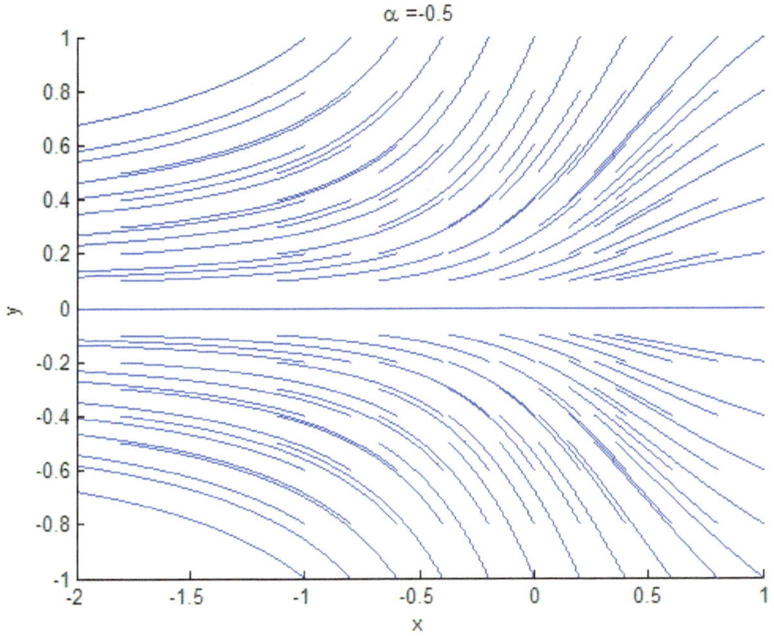

Phase portrait showing saddle-node bifurcation.

Bifurcation Types

It is useful to divide bifurcations into two principal classes:

- Local bifurcations, which can be analysed entirely through changes in the local stability properties of equilibria, periodic orbits or other invariant sets as parameters cross through critical thresholds; and
- Global bifurcations, which often occur when larger invariant sets of the system 'collide' with each other, or with equilibria of the system. They cannot be detected purely by a stability analysis of the equilibria (fixed points).

Local Bifurcations

A local bifurcation occurs when a parameter change causes the stability of an equilibrium (or fixed point) to change. In continuous systems, this corresponds to the real part of an eigenvalue of an equilibrium passing through zero. In discrete systems (those described by maps rather than ODEs), this corresponds to a fixed point having a Floquet multiplier with modulus equal to one. In both cases, the equilibrium is non-hyperbolic at the bifurcation point. The topological changes in the phase portrait of the system can be confined to arbitrarily small neighbourhoods of the bifurcating fixed points by moving the bifurcation parameter close to the bifurcation point (hence ‹local').

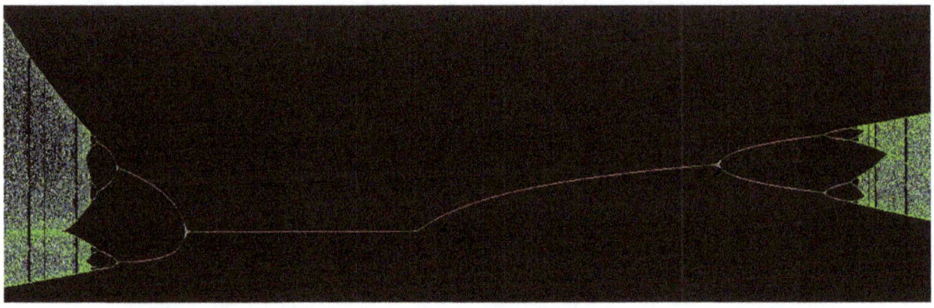

Period-halving bifurcations (L) leading to order, followed by period doubling bifurcations (R) leading to chaos.

More technically, consider the continuous dynamical system described by the ODE

$$\dot{x} = f(x, \lambda) \quad f\colon \mathbb{R}^n \times \mathbb{R} \to \mathbb{R}^n.$$

A local bifurcation occurs at (x_0, λ_0) if the Jacobian matrix $\mathrm{d}f_{x_0,\lambda_0}$ has an eigenvalue with zero real part. If the eigenvalue is equal to zero, the bifurcation is a steady state bifurcation, but if the eigenvalue is non-zero but purely imaginary, this is a Hopf bifurcation.

For discrete dynamical systems, consider the system

$$x_{n+1} = f(x_n, \lambda).$$

Then a local bifurcation occurs at (x_0, λ_0) if the matrix $\mathrm{d}f_{x_0,\lambda_0}$ has an eigenvalue with modulus equal to one. If the eigenvalue is equal to one, the bifurcation is either a saddle-node (often called fold bifurcation in maps), transcritical or pitchfork bifurcation. If the eigenvalue is equal to −1, it is a period-doubling (or flip) bifurcation, and otherwise, it is a Hopf bifurcation.

Examples of local bifurcations include:

- Saddle-node (fold) bifurcation
- Transcritical bifurcation
- Pitchfork bifurcation
- Period-doubling (flip) bifurcation
- Hopf bifurcation
- Neimark–Sacker (secondary Hopf) bifurcation

Global Bifurcations

Global bifurcations occur when 'larger' invariant sets, such as periodic orbits, collide with equilibria. This causes changes in the topology of the trajectories in the phase space which cannot be confined to a small neighbourhood, as is the case with local bifurcations. In fact, the changes in topology extend out to an arbitrarily large distance (hence 'global').

Examples of global bifurcations include:

- Homoclinic bifurcation in which a limit cycle collides with a saddle point.

- Heteroclinic bifurcation in which a limit cycle collides with two or more saddle points.
- Infinite-period bifurcation in which a stable node and saddle point simultaneously occur on a limit cycle.
- Blue sky catastrophe in which a limit cycle collides with a nonhyperbolic cycle.

Global bifurcations can also involve more complicated sets such as chaotic attractors (e.g. crises).

Codimension of a Bifurcation

The codimension of a bifurcation is the number of parameters which must be varied for the bifurcation to occur. This corresponds to the codimension of the parameter set for which the bifurcation occurs within the full space of parameters. Saddle-node bifurcations and Hopf bifurcations are the only generic local bifurcations which are really codimension-one (the others all having higher codimension). However, transcritical and pitchfork bifurcations are also often thought of as codimension-one, because the normal forms can be written with only one parameter.

An example of a well-studied codimension-two bifurcation is the Bogdanov–Takens bifurcation.

Applications in Semiclassical and Quantum Physics

Bifurcation theory has been applied to connect quantum systems to the dynamics of their classical analogues in atomic systems, molecular systems, and resonant tunneling diodes. Bifurcation theory has also been applied to the study of laser dynamics and a number of theoretical examples which are difficult to access experimentally such as the kicked top and coupled quantum wells. The dominant reason for the link between quantum systems and bifurcations in the classical equations of motion is that at bifurcations, the signature of classical orbits becomes large, as Martin Gutzwiller points out in his classic work on quantum chaos. Many kinds of bifurcations have been studied with regard to links between classical and quantum dynamics including saddle node bifurcations, Hopf bifurcations, umbilic bifurcations, period doubling bifurcations, reconnection bifurcations, tangent bifurcations, and cusp bifurcations.

REFERENCES

1. Bellman, Richard; Cooke, Kenneth L. (1963). Differential-difference equations. New York-London: Academic Press. ISBN 978-0-12-084850-8.
2. Driver, Rodney D. (1977). Ordinary and Delay Differential Equations. New York: Springer Verlag. ISBN 0-387-90231-7.
3. Michiels, Wim; Niculescu, Silviu-Iulian (2007). Stability and stabilization of time-delay systems. An eigenvalue based approach. doi:10.1137/1.9780898718645. ISBN 978-0-89871-632-0.

4. Briat, Corentin (2015). Linear Parameter-Varying and Time-Delay Systems. Analysis, Observation, Filtering & Control. Springer Verlag Heidelberg. ISBN 978-3-662-44049-0.
5. Kolmanovskii, V.; Myshkis, A. (1992). Applied Theory of Functional Differential Equations. The Netherlands: Kluwer Academic Publishers. ISBN 0-7923-2013-1.
6. Hale, Jack K. (1971). Functional Differential Equation. United States: Springer-Verlag. ISBN 0-387-90023-3.
7. Hale, Jack K.; Verduyn Lunel, Sjoerd M. (1993). Introduction to Functional Differential Equations. United States: Springer-Verlag. ISBN 0-387-94076-6.
8. Falbo, Clement E. "Some Elementary Methods for Solving Functional Differential Equations" (PDF). Somona State University.
9. Guo, S.; Wu, J. (2013). Bifurcation Theory of Functional Differential Equations. New York: Springer. pp. 41–60. ISBN 978-1-4614-6991-9.
10. Bellman, Richard; Cooke, Kenneth L. (1963). Differential-Difference Equations. New York, NY: Academic Press. pp. 42–49. ISBN 978-0124109735.
11. Barnes, B.; Fulford, G. R. (2015). Mathematical Modelling with Case Studies. Taylor & Francis Group LLC. pp. 75–77. ISBN 978-1-4822-4772-5.
12. Schmitt, Klaus, ed. (1972). Delay and Functional Differential Equations and Their Applications. United States: Academic Press.
13. Xu, Changjin; Wu, Yusen (2013). "Dynamics in a Lotka–Volterra Predator–Prey Model with Time-varying Delays". Abstract and Applied Analysis. 2013.
14. Traian Lalescu, Introduction à la théorie des équations intégrales. Avec une préface de É. Picard, Paris: A. Hermann et Fils, 1912. VII + 152 pp.
15. Hazewinkel, Michiel, ed. (2001), "Volterra equation", Encyclopedia of Mathematics, Springer, ISBN 978-1-55608-010-4
16. Weisstein, Eric W. "Volterra Integral Equation of the First Kind". MathWorld.
17. Weisstein, Eric W. "Volterra Integral Equation of the Second Kind". MathWorld.
18. Integral Equations: Exact Solutions at EqWorld: The World of Mathematical Equations
19. Press, WH; Teukolsky, SA; Vetterling, WT; Flannery, BP (2007). "Section 19.2. Volterra Equations". Numerical Recipes: The Art of Scientific Computing (3rd ed.). New York: Cambridge University Press. ISBN 978-0-521-88068-8.
20. E. R. Leigh (1968) The ecological role of Volterra's equations, in Some Mathematical Problems in Biology – a modern discussion using Hudson's Bay Company data on lynx and hares in Canada from 1847 to 1903.
21. Understanding Nonlinear Dynamics. Daniel Kaplan and Leon Glass.
22. J.D. Murray. Mathematical Biology I: An Introduction. Springer-Verlag, 2003
23. James A. Yorke; William N. Anderson, Jr. Predator-Prey Patterns (Volterra-Lotka equations). PNAS, vol. 70, n° 7, pp. 2069-2071, July 1973

24. J. Llibre and C. Valls (2007). " Global analytic first integrals for the real planar Lotka-Volterra system", J. Math. Phys. 48, 033507 (2007); doi:10.1063/1.2713076

25. Blanchard, P.; Devaney, R. L.; Hall, G. R. (2006). Differential Equations. London: Thompson. pp. 96–111. ISBN 0-495-01265-3.

26. Henri Poincaré. "L'Équilibre d'une masse fluide animée d'un mouvement de rotation". Acta Mathematica, vol.7, pp. 259-380, Sept 1885.

27. Gao, J.; Delos, J. B. (1997). "Quantum manifestations of bifurcations of closed orbits in the photoabsorption spectra of atoms in electric fields". Phys. Rev. A. 56 (1): 356–364. Bibcode:1997PhRvA..56..356G. doi:10.1103/PhysRevA.56.356.

28. Peters, A. D.; Jaffé, C.; Delos, J. B. (1994). "Quantum Manifestations of Bifurcations of Classical Orbits: An Exactly Solvable Model". Phys. Rev. Lett. 73 (21): 2825–2828. Bibcode:1994PhRvL..73.2825P. doi:10.1103/PhysRevLett.73.2825. PMID 10057205.

29. Courtney, Michael; Jiao, Hong; Spellmeyer, Neal; Kleppner, Daniel; Gao, J.; Delos, J. B.; et al. (1995). "Closed Orbit Bifurcations in Continuum Stark Spectra". Phys. Rev. Lett. 74 (9): 1538–1541. Bibcode:1995PhRvL..74.1538C. doi:10.1103/PhysRevLett.74.1538. PMID 10059054.

30. Founargiotakis, M.; Farantos, S. C.; Skokos, Ch.; Contopoulos, G. (1997). "Bifurcation diagrams of periodic orbits for unbound molecular systems: FH2". Chemical Physics Letters. 277 (5–6): 456–464. Bibcode:1997CPL...277..456F. doi:10.1016/S0009-2614(97)00931-7.

31. Monteiro, T. S. & Saraga, D. S. (2001). "Quantum Wells in Tilted Fields:Semiclassical Amplitudes and Phase Coherence Times". Foundations of Physics. 31 (2): 355–370. doi:10.1023/A:1017546721313.

32. Wieczorek, S.; Krauskopf, B.; Simpson, T. B. & Lenstra, D. (2005). "The dynamical complexity of optically injected semiconductor lasers". Physics Reports. 416 (1–2): 1–128. Bibcode:2005PhR...416....1W. doi:10.1016/j.physrep.2005.06.003.

33. Stamatiou, G. & Ghikas, D. P. K. (2007). "Quantum entanglement dependence on bifurcations and scars in non-autonomous systems. The case of quantum kicked top". Physics Letters A. 368 (3–4): 206–214. arXiv:quant-ph/0702172Freely accessible. Bibcode:2007PhLA..368..206S. doi:10.1016/j.physleta.2007.04.003.

34. Galan, J.; Freire, E. (1999). "Chaos in a Mean Field Model of Coupled Quantum Wells; Bifurcations of Periodic Orbits in a Symmetric Hamiltonian System". Reports on Mathematical Physics. 44 (1–2): 87–94. Bibcode:1999RpMP...44...87G. doi:10.1016/S0034-4877(99)80148-7.

35. Kleppner, D.; Delos, J. B. (2001). "Beyond quantum mechanics: Insights from the work of Martin Gutzwiller". Foundations of Physics. 31 (4): 593–612. doi:10.1023/A:1017512925106.

36. Gutzwiller, Martin C. (1990). Chaos in Classical and Quantum Mechanics. New York: Springer-Verlag. ISBN 0-387-97173-4.

CHAPTER 7

Autonomous Differential Equations

Chapter Contents

- **Introduction**
- **Definition**
- **Properties**
- **Time-Invariant System**
- **Non-Autonomous System**

An autonomous differential equation is an equation of the form

$$\frac{dy}{dt} = f(y).$$

Let's think of t as indicating time. This equation says that the rate of change dy/dt of the function y(t) is given by a some rule. The rule says that if the current value is y, then the rate of change is f(y).

The equation is called a differential equation, because it is an equation involving the derivative dy/dt. The differential equation is called autonomous because the rule doesn't care what time t it is. It only cares about the current value of the variable y.

Given an autonomous differential equation, we'll often want to solve the equation, which means find a function a y(t) whose derivative dy/dt is equal to f(y)).

INTRODUCTION

In mathematics, an autonomous system or autonomous differential equation is a system of ordinary differential equations which does not explicitly depend on the independent variable. When the variable is time, they are also called time-invariant systems.

Many laws in physics, where the independent variable is usually assumed to be time, are expressed as autonomous systems because it is assumed the laws of nature which hold now are identical to those for any point in the past or future.

Autonomous systems are closely related to dynamical systems. Any autonomous system can be transformed into a dynamical system and, using very weak assumptions, a dynamical system can be transformed into an autonomous system.

DEFINITION

An autonomous system is a system of ordinary differential equations of the form

$$\frac{d}{dt}x(t) = f(x(t))$$

where x takes values in n-dimensional Euclidean space and t is usually time.

It is distinguished from systems of differential equations of the form

$$\frac{d}{dt}x(t) = g(x(t), t)$$

in which the law governing the rate of motion of a particle depends not only on the particle's location, but also on time; such systems are not autonomous.

PROPERTIES

Let $x_1(t)$ be a unique solution of the initial value problem for an autonomous system

$$\frac{d}{dt}x(t) = f(x(t)), \quad x(0) = x_0$$

Then $x_2(t) = x_1(t - t_0)$ solves

$$\frac{d}{dt}x(t) = f(x(t)), \quad x(t_0) = x_0$$

Indeed, denoting $s = t - t_0$ we have $x_1(s) = x_2(t)$ and $ds = dt$, thus

$$\frac{d}{dt}x_2(t) = \frac{d}{dt}x_1(t - t_0) = \frac{d}{ds}x_1(s) = f(x_1(s)) = f(x_2(t)).$$

For the initial condition, the verification is trivial,

$$x_2(t_0) = x_1(t_0 - t_0) = x_1(0) = x_0.$$

Example

The equation $y' = (2 - y)y$ is autonomous, since the independent variable, let us call it x, does not explicitly appear in the equation. To plot the slope field and isocline for this equation, one can use the following code in GNU

```
Ffun = @(X,Y)(2-Y).*Y;              % function f(x,y)=(2-y)y
[X,Y]=meshgrid(0:.2:6,-1:.2:3);     % choose the plot sizes
DY=Ffun(X,Y); DX=ones(size(DY));    % generate the plot values
quiver(X,Y,DX,DY, 'k');             % plot the direction field in black
hold on;
contour(X,Y,DY,[0 1 2], 'g');       % add the isoclines(0 1 2) in green
title('Slope field and isoclines for f(x,y)=(2-y)y')
```

One can observe from the plot that the function $(2-y)y$ is x -invariant, and so is the shape of the solution, i.e. $y(x) = y(x - x_0)$.

Solving the equation symbolically in MATLAB, by running

```
y=dsolve('Dy=(2-y)*y','x'); % solve the equation symbolically
```

we obtain two equilibrium solutions, $y = 0$ and $y = 2$, and a third solution involving an unknown constant C_3,

```
y(3)=-2/(exp(C3 - 2*x) - 1)
```

Picking up some specific values for the initial condition, we can add the plot of several solutions

```
y1=dsolve('Dy=(2-y)*y','y(1)=1','x'); % solve the initial value problem symbolically
y2=dsolve('Dy=(2-y)*y','y(2)=1','x'); % for different initial conditions
y3=dsolve('Dy=(2-y)*y','y(3)=1','x'); y4=dsolve('Dy=(2-y)*y','y(1)=3','x');
y5=dsolve('Dy=(2-y)*y','y(2)=3','x'); y6=dsolve('Dy=(2-y)*y','y(3)=3','x');
ezplot(y1, [0 6]); ezplot(y2, [0 6]); % plot the solutions
ezplot(y3, [0 6]); ezplot(y4, [0 6]); ezplot(y5, [0 6]); ezplot(y6, [0 6]);
title('Slope field, isoclines and solutions for f(x,y)=(2-y)y')
legend('Slope field', 'Isoclines', 'Solutions y_{1..6}');
text([1 2 3], [1 1 1], strcat('\leftarrow', {'y_1','y_2', 'y_3'}));
text([1 2 3], [3 3 3], strcat('\leftarrow', {'y_4','y_5', 'y_6'}));
grid on;
```

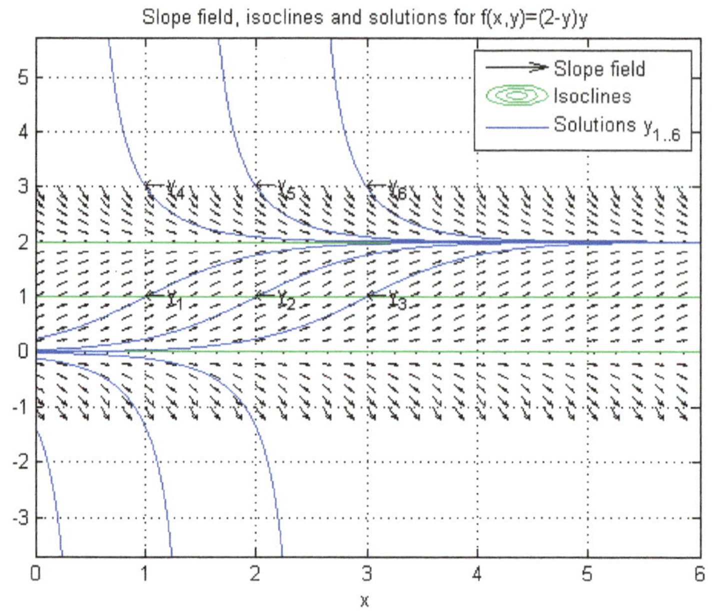

Slope field with isoclines and solutions

Qualitative Analysis

Autonomous systems can be analyzed qualitatively using the phase space; in the one-variable case, this is the phase line.

Solution Techniques

The following techniques apply to one-dimensional autonomous differential equations. Any one-dimensional equation of order n is equivalent to an n-dimensional first-order system (as described in Ordinary differential equation#Reduction to a first order system), but not necessarily vice versa.

First Order

The first-order autonomous equation

$$\frac{dx}{dt} = f(x)$$

is separable, so it can easily be solved by rearranging it into the integral form

$$t + C = \int \frac{dx}{f(x)}$$

Second Order

The second-order autonomous equation

$$\frac{d^2x}{dt^2} = f(x, x')$$

is more difficult, but it can be solved by introducing the new variable

$$v = \frac{dx}{dt}$$

and expressing the second derivative of x (via the chain rule) as

$$\frac{d^2x}{dt^2} = \frac{dv}{dt} = \frac{dx}{dt}\frac{dv}{dx} = v\frac{dv}{dx}$$

so that the original equation becomes

$$v\frac{dv}{dx} = f(x, v)$$

which is a first order equation containing no reference to the independent variable t and if solved provides vas a function of x. Then, recalling the definition of v:

$$\frac{dx}{dt} = v(x) \quad \Rightarrow \quad t + C = \int \frac{dx}{v(x)}$$

which is an implicit solution.

Special Case: x'' = f(x)

The special case where f is independent of x'

$$\frac{d^2x}{dt^2} = f(x)$$

benefits from separate treatment. These types of equations are very common in classical mechanics because they are always Hamiltonian systems.

The idea is to make use of the identity (barring division by zero issues)

$$\frac{dx}{dt} = \left(\frac{dt}{dx}\right)^{-1}$$

which follows from the chain rule. Note aside then that by inverting both sides of a first order autonomous system, one can immediately integrate with respect to x:

$$\frac{dx}{dt} = f(x) \quad \Rightarrow \quad \frac{dt}{dx} = \frac{1}{f(x)} \quad \Rightarrow \quad t + C = \int \frac{dx}{f(x)}$$

which is another way to view the separation of variables technique. A natural question is then: can we do something like this with higher order equations? The answer is yes for second order equations, but there's more work to do. The second derivative must be expressed as a derivative with respect to x instead of t:

$$\frac{d^2x}{dt^2} = \frac{d}{dt}\left(\frac{dx}{dt}\right) = \frac{d}{dx}\left(\frac{dx}{dt}\right)\frac{dx}{dt} = \frac{d}{dx}\left(\left(\frac{dt}{dx}\right)^{-1}\right)\left(\frac{dt}{dx}\right)^{-1} =$$
$$= -\left(\frac{dt}{dx}\right)^{-2}\frac{d^2t}{dx^2}\left(\frac{dt}{dx}\right)^{-1} = -\left(\frac{dt}{dx}\right)^{-3}\frac{d^2t}{dx^2} =$$
$$= \frac{d}{dx}\left(\frac{1}{2}\left(\frac{dt}{dx}\right)^{-2}\right)$$

To reemphasize: what's been accomplished is that the second derivative in t has been expressed as a derivative in x. The original second order equation may then finally be integrated:

$$\frac{d^2x}{dt^2} = f(x)$$
$$\frac{d}{dx}\left(\frac{1}{2}\left(\frac{dt}{dx}\right)^{-2}\right) = f(x)$$
$$\left(\frac{dt}{dx}\right)^{-2} = 2\int f(x)dx + C_1$$
$$\frac{dt}{dx} = \pm\frac{1}{\sqrt{2\int f(x)dx + C_1}}$$
$$t + C_2 = \pm\int \frac{dx}{\sqrt{2\int f(x)dx + C_1}}$$

This is an implicit solution, and beyond that the greatest potential problem is inability to simplify the integrals, which implies difficulty or impossibility in evaluating the integration constants.

Special Case: x'' = x'ⁿ f(x)

Using the above mentality, we can extend the technique to the more general equation

$$\frac{d^2x}{dt^2} = \left(\frac{dx}{dt}\right)^n f(x)$$

Where n is some parameter not equal to two. This will work since the second derivative can be written in a form involving a power of x'. Rewriting the second derivative, rearranging, and expressing the left side as a derivative:

$$-\left(\frac{dt}{dx}\right)^{-3}\frac{d^2t}{dx^2}=\left(\frac{dt}{dx}\right)^{-n}f(x)$$

$$-\left(\frac{dt}{dx}\right)^{n-3}\frac{d^2t}{dx^2}=f(x)$$

$$\frac{d}{dx}\left(\frac{1}{2-n}\left(\frac{dt}{dx}\right)^{n-2}\right)=f(x)$$

$$\left(\frac{dt}{dx}\right)^{n-2}=(2-n)\int f(x)dx+C_1$$

$$t+C_2=\int\left((2-n)\int f(x)dx+C_1\right)^{\frac{1}{n-2}}dx$$

The right will carry +/- if n is even. The treatment must be different if n=2:

$$-\left(\frac{dt}{dx}\right)^{-1}\frac{d^2t}{dx^2}=f(x)$$

$$-\frac{d}{dx}\left(\ln\left(\frac{dt}{dx}\right)\right)=f(x)$$

$$\frac{dt}{dx}=C_1e^{-\int f(x)dx}$$

$$t+C_2=C_1\int e^{-\int f(x)dx}dx$$

Higher Orders

There is no analogous method for solving third- or higher-order autonomous equations. Such equations can only be solved exactly if they happen to have some other simplifying property, for instance linearity or dependence of the right side of the equation on the dependent variable only (i.e., not its derivatives). This should not be surprising, considering that nonlinear autonomous systems in three dimensions can produce truly chaotic behavior such as the Lorenz attractor and the Rössler attractor.

With this mentality, it also isn't too surprising that general non-autonomous equations of second order can't be solved explicitly, since these can also be chaotic (an example of this is a periodically forced pendulum).

TIME-INVARIANT SYSTEM

A time-invariant (TIV) system is a system whose output does not depend explicitly on time. Such systems are regarded as a class of systems in the field of system analysis. Lack of time dependence is captured in the following mathematical property of such a system:

If the input signal x(t) produces an output y(t) then any time shifted input, x (t+δ), results in a time-shifted output y (t+ δ)

This property can be satisfied if the transfer function of the system is not a function of time except expressed by the input and output.

In the context of a system schematic, this property can also be stated as follows:

If a system is time-invariant then the system block commutes with an arbitrary delay.

If a time-invariant system is also linear, it is the subject of LTI system theory (linear time-invariant) with direct applications in NMR spectroscopy, seismology, circuits, signal processing, control theory, and other technical areas. Nonlinear time-invariant systems lack a comprehensive, governing theory. Discrete time-invariant systems are known as shift-invariant systems. Systems which lack the time-invariant property are studied as time-variant systems.

Simple Example

To demonstrate how to determine if a system is time-invariant, consider the two systems:

System A: $y(t) = t\,x(t)$

System B: $y(t) = 10x(t)$

Since system A explicitly depends on *t* outside of $x(t)$ and $y(t)$, it is not time-invariant. System B, however, does not depend explicitly on *t* so it is time-invariant.

Formal Example

A more formal proof of why systems A and B above differ is now presented. To perform this proof, the second definition will be used.

System A:

Start with a delay of the input $x_d(t) = x(t+\delta)$

$$y(t) = t\,x(t)$$
$$y_1(t) = t\,x_d(t) = t\,x(t+\delta)$$

Now delay the output by δ

$$y(t) = t\,x(t)$$
$$y_2(t) = y(t+\delta) = (t+\delta)x(t+\delta)$$

Clearly $y_1(t) \neq y_2(t)$, therefore the system is not time-invariant.

System B:

Start with a delay of the input $x_d(t) = x(t+\delta)$

$$y(t) = 10\,x(t)$$
$$y_1(t) = 10\,x_d(t) = 10\,x(t+\delta)$$

Now delay the output by δ

$$y(t) = 10\,x(t)$$
$$y_2(t) = y(t+\delta) = 10\,x(t+\delta)$$

Clearly $y_1(t) = y_2(t)$, therefore the system is time-invariant.

Abstract Example

We can denote the shift operator by $\mathbb{T}_r$ where r is the amount by which a vector's index set should be shifted. For example, the "advance-by-1" system

$$x(t+1) = \delta(t+1) * x(t)$$

can be represented in this abstract notation by

$$\tilde{x}_1 = \mathbb{T}_1 \, \tilde{x}$$

where $\tilde{x}$ is a function given by

$$\tilde{x} = x(t) \, \forall \, t \in \mathbb{R}$$

with the system yielding the shifted output

$$\tilde{x}_1 = x(t+1) \, \forall \, t \in \mathbb{R}$$

So $\mathbb{T}_1$ is an operator that advances the input vector by 1.

Suppose we represent a system by an operator $\mathbb{H}$. This system is time-invariant if it commutes with the shift operator, i.e.,

$$\mathbb{T}_r \, \mathbb{H} = \mathbb{H} \, \mathbb{T}_r \;\; \forall \, r$$

If our system equation is given by

$$\tilde{y} = \mathbb{H} \, \tilde{x}$$

then it is time-invariant if we can apply the system operator $\mathbb{H}$ on $\tilde{x}$ followed by the shift operator $\mathbb{T}_r$ or we can apply the shift operator $\mathbb{T}_r$ followed by the system operator $\mathbb{H}$, with the two computations yielding equivalent results.

Applying the system operator first gives

Applying the system operator first gives

$$\mathbb{T}_r \, \mathbb{H} \, \tilde{x} = \mathbb{T}_r \, \tilde{y} = \tilde{y}_r$$

Applying the shift operator first gives

$$\mathbb{H} \, \mathbb{T}_r \, \tilde{x} = \mathbb{H} \, \tilde{x}_r$$

If the system is time-invariant, then

$$\mathbb{H} \, \tilde{x}_r = \tilde{y}_r$$

NON-AUTONOMOUS SYSTEM

In mathematics, an autonomous system is a dynamic equation on a smooth manifold. A non-autonomous system is a dynamic equation on a smooth fiber bundle $Q \to \mathbb{R}$ over $\mathbb{R}$. For instance, this is the case of non-autonomous mechanics.

An *r*-order differential equation on a fiber bundle $Q \to \mathbb{R}$ is represented by a closed subbundle of a jet bundle $J^rQ$of $Q \to \mathbb{R}$. A dynamic equation on $Q \to \mathbb{R}$is a differential equation which is algebraically solved for a higher-order derivatives.

In particular, a first-order dynamic equation on a fiber bundle $Q \to \mathbb{R}$ is a kernel of the covariant differential of some connection Γ on $Q \to \mathbb{R}$. Given bundle coordinates (t,q^i) on Q and the adapted coordinates (t,q^i,q^i_t) on a first-order jet manifold J^1Q, a first-order dynamic equation reads

$$q^i_t = \Gamma(t,q^i).$$

For instance, this is the case of Hamiltonian non-autonomous mechanics.

A second-order dynamic equation

$$q^i_{tt} = \xi^i(t,q^j,q^j_t)$$

on Q → R is defined as a holonomic connection ξ on a jet bundle J^1Q → R. This equation also is represented by a connection on an affine jet bundle J^1Q → Q. Due to the canonical imbedding J^1Q → T Q, it is equivalent to a geodesic equation on the tangent bundle TQ of Q . A free motion equation in non-autonomous mechanics exemplifies a second-order non-autonomous dynamic equation.

REFERENCES

1. Boyce, William E.; Richard C. DiPrima (2005). Elementary Differential Equations and Boundary Volume Problems (8th ed.). John Wiley & Sons. p. 133. ISBN 0-471-43338-1.
2. Second order autonomous equation at eqworld.
3. Third order autonomous equation at eqworld.
4. Fourth order autonomous equation at eqworld.
5. Blanchard; Devaney; Hall (2005). Differential Equations. Brooks/Cole Publishing Co. pp. 540–543. ISBN 0-495-01265-3.
6. De Leon, M., Rodrigues, P., Methods of Differential Geometry in Analytical Mechanics (North Holland, 1989).
7. Giachetta, G., Mangiarotti, L., Sardanashvily, G., Geometric Formulation of Classical and Quantum Mechanics (World Scientific, 2010) ISBN 981-4313-72-6 (arXiv: 0911.0411).

CHAPTER 8

Stochastic Differential Equations

Chapter Contents

- **Introduction**
- **Background**
- **Terminology**
- **Stochastic Partial Differential Equation**
- **Autoregressive Model**

Stochastic differential equations (SDEs) occur where a system described by differential equations is influenced by random noise. Stochastic differential equations are used in finance (interest rate, stock prices, ...), biology (population, epidemics, ...

INTRODUCTION

A stochastic differential equation (SDE) is a differential equation in which one or more of the terms is a stochastic process, resulting in a solution which is also a stochastic process. SDEs are used to model various phenomena such as unstable stock prices or physical systems subject to thermal fluctuations. Typically, SDEs contain a variable which represents random white noise calculated as the derivative of Brownian motion or the Wiener process. However, other types of random behaviour are possible, such as jump processes.

BACKGROUND

Early work on SDEs was done to describe Brownian motion in Einstein's famous paper, and at the same time by Smoluchowski. However, one of the earlier works related to Brownian motion is credited to Bachelier (1900) in his thesis 'Theory of Speculation'.

This work was followed upon by Langevin. Later Itô and Stratonovich put SDEs on more solid mathematical footing.

TERMINOLOGY

In physical science, SDEs are usually written as Langevin equations. These are sometimes ambiguously called "the Langevin equation" even though there are many possible forms. Those forms consist of an ordinary differential equation containing a deterministic function and an additional random white noise term. A second form includes the Smoluchowski equation or the Fokker-Planck equation. These are partial differential equations which describe the time evolution of probability distribution functions. The third form is the Itô stochastic differential equation, which is most frequently used in mathematics and quantitative finance. This is similar to the Langevin form, but it is usually written in differential notation. SDEs are denoted in two varieties, corresponding to two versions of stochastic calculus.

Stochastic Calculus

Brownian motion or the Wiener process was discovered to be exceptionally complex mathematically. The Wiener process is almost surely nowhere differentiable; thus, it requires its own rules of calculus. There are two dominating versions of stochastic calculus, the Itô stochastic calculus and the Stratonovich stochastic calculus. Each of the two has advantages and disadvantages, and newcomers are often confused whether the one is more appropriate than the other in a given situation. Guidelines exist (e.g. Øksendal, 2003) and conveniently, one can readily convert an Itô SDE to an equivalent Stratonovich SDE and back again. Still, one must be careful which calculus to use when the SDE is initially written down.

Numerical Solutions

Numerical solution of stochastic differential equations and especially stochastic partial differential equations is a young field relatively speaking. Almost all algorithms that are used for the solution of ordinary differential equations will work very poorly for SDEs, having very poor numerical convergence. A textbook describing many different algorithms is Kloeden & Platen (1995).

Methods include the Euler–Maruyama method, Milstein method and Runge–Kutta method (SDE).

Use in Physics

In physics, SDEs are typically written in the Langevin form and referred to as "the Langevin equation." For example, a general coupled set of first-order SDEs is often written in the form:

$$\dot{x}_i = \frac{dx_i}{dt} = f_i(\mathbf{x}) + \sum_{m=1}^{n} g_i^m(\mathbf{x})\eta_m(t),$$

where $\mathbf{x} = \{x_i | 1 \le i \le k\}$ is the set of unknowns, the f_i and g_i are arbitrary functions and the η_m are random functions of time, often referred to as "noise terms". This form is usually usable because there are standard techniques for transforming higher-order equations into several coupled first-order equations by introducing new unknowns. If the g_i are constants, the system is said to be subject to additive noise, otherwise it is said to be subject to multiplicative noise. This term is somewhat misleading as it has come to mean the general case even though it appears to imply the limited case in which $g(x) \propto x$. Additive noise is the simpler of the two cases; in that situation the Langevin equation has only one natural notion of solution and, in the linear case, this solution has an explicit expression found using the ordinary rules of calculus as if the η_m were functions. However, in the case of multiplicative noise, the Langevin equation is not a well-defined entity on its own, and it must be specified whether the Langevin equation should be interpreted as an Itô SDE or a Stratonovich SDE.

In physics, the main method of solution is to find the probability distribution function as a function of time using the equivalent Fokker-Planck equation (FPE). The Fokker-Planck equation is a deterministic partial differential equation. It tells how the probability distribution function evolves in time similarly to how the Schrödinger equation gives the time evolution of the quantum wave function or the diffusion equation gives the time evolution of chemical concentration. Alternatively numerical solutions can be obtained by Monte Carlo simulation. Other techniques include the path integration that draws on the analogy between statistical physics and quantum mechanics (for example, the Fokker-Planck equation can be transformed into the Schrödinger equation by rescaling a few variables) or by writing down ordinary differential equations for the statistical moments of the probability distribution function.

Use in Probability and Mathematical Finance

The notation used in probability theory (and in many applications of probability theory, for instance mathematical finance) is slightly different. This notation makes the exotic nature of the random function of time η_m in the physics formulation more explicit. It is also the notation used in publications on numerical methods for solving stochastic differential equations. In strict mathematical terms, η_m cannot be chosen as an ordinary function, but only as a generalized function. The mathematical formulation treats this complication with less ambiguity than the physics formulation.

A typical equation is of the form

$$\mathrm{d}X_t = \mu(X_t, t)\,\mathrm{d}t + \sigma(X_t, t)\,\mathrm{d}B_t,$$

where B denotes a Wiener process (Standard Brownian motion). This equation should be interpreted as an informal way of expressing the corresponding integral equation

$$X_{t+s} - X_t = \int_t^{t+s} \mu(X_u, u)du + \int_t^{t+s} \sigma(X_u, u)\, dB_u.$$

The equation above characterizes the behavior of the continuous time stochastic process X_t as the sum of an ordinary Lebesgue integral and an Itô integral. A heuristic (but very helpful) interpretation of the stochastic differential equation is that in a small time interval of length δ the stochastic process X_t changes its value by an amount that is normally distributed with expectation $\mu(X_t, t)\ \delta$ and variance $\sigma(X_t, t)^2\ \delta$ and is independent of the past behavior of the process. This is so because the increments of a Wiener process are independent and normally distributed. The function μ is referred to as the drift coefficient, while σ is called the diffusion coefficient. The stochastic process X_t is called a diffusion process, and is usually a Markov process.

The formal interpretation of an SDE is given in terms of what constitutes a solution to the SDE. There are two main definitions of a solution to an SDE, a strong solution and a weak solution. Both require the existence of a process X_t that solves the integral equation version of the SDE. The difference between the two lies in the underlying probability space (Ω, F, Pr). A weak solution consists of a probability space and a process that satisfies the integral equation, while a strong solution is a process that satisfies the equation and is defined on a given probability space.

An important example is the equation for geometric Brownian motion

$$dX_t = \mu X_t\, dt + \sigma X_t\, dB_t.$$

which is the equation for the dynamics of the price of a stock in the Black Scholes options pricing model of financial mathematics.

There are also more general stochastic differential equations where the coefficients μ and σ depend not only on the present value of the process X_t, but also on previous values of the process and possibly on present or previous values of other processes too. In that case the solution process, X, is not a Markov process, and it is called an Itô process and not a diffusion process. When the coefficients depends only on present and past values of X, the defining equation is called a stochastic delay differential equation.

Existence and Uniqueness of Solutions

As with deterministic ordinary and partial differential equations, it is important to know whether a given SDE has a solution, and whether or not it is unique. The following is a typical existence and uniqueness theorem for Itô SDEs taking values in n-dimensional Euclidean space $\mathbf{R}^n$ and driven by an m-dimensional Brownian motion B; the proof may be found in Øksendal (2003, §5.2).

Let T > 0, and let

$$\mu : \mathbb{R}^n \times [0,T] \to \mathbb{R}^n;$$
$$\sigma : \mathbb{R}^n \times [0,T] \to \mathbb{R}^{n \times m};$$

be measurable functions for which there exist constants C and D such that

$$|\mu(x,t)| + |\sigma(x,t)| \le C(1+|x|);$$
$$|\mu(x,t) - \mu(y,t)| + |\sigma(x,t) - \sigma(y,t)| \le D|x-y|;$$

for all $t \in [0, T]$ and all x and $y \in \mathbf{R}^n$, where

$$|\sigma|^2 = \sum_{i,j=1}^{n} |\sigma_{ij}|^2.$$

Let Z be a random variable that is independent of the σ-algebra generated by B_s, s ≥ 0, and with finite second moment:

$$\mathbb{E}\left[|Z|^2\right] < +\infty.$$

Then the stochastic differential equation/initial value problem

$$\mathrm{d}X_t = \mu(X_t,t)\,\mathrm{d}t + \sigma(X_t,t)\,\mathrm{d}B_t \text{ for } t \in [0,T];$$
$$X_0 = Z;$$

has a Pr-almost surely unique t-continuous solution (t, ω) |→ $X_t(\omega)$ such that X is adapted to the filtration F_t^Z generated by Z and B_s, s ≤ t, and

$$\mathbb{E}\left[\int_0^T |X_t|^2\,\mathrm{d}t\right] < +\infty.$$

Some Explicitly Solvable SDEs

Linear SDE: General Case

$$dX_t = (a(t)X_t + c(t))dt + (b(t)X_t + d(t))dW_t$$
$$X_t = \Phi_{t,t_0}\left(X_{t_0} + \int_{t_0}^{t} \Phi_{s,t_0}^{-1}(c(s) - b(s)d(s))ds + \int_{t_0}^{t} \Phi_{s,t_0}^{-1} d(s)dW_s\right)$$

where

$$\Phi_{t,t_0} = \exp\left(\int_{t_0}^{t}\left(a(s) - \frac{b^2(s)}{2}\right)ds + \int_{t_0}^{t} b(s)dW_s\right)$$

Reducible SDEs: Case 1

$$dX_t = \frac{1}{2}f(X_t)f'(X_t)dt + f(X_t)W_t$$

for a given differentiable function f is equivalent to the Stratonovich SDE

$$dX_t = f(X_t) \circ W_t$$

which has a general solution

$$X_t = h^{-1}(W_t + h(X_0))$$

where

$$h(x) = \int^{x} \frac{ds}{f(s)}$$

Reducible SDEs: Case 2

$$dX_t = \left(\alpha f(X_t) + \frac{1}{2} f(X_t) f'(X_t)\right) dt + f(X_t) W_t$$

for a given differentiable function f is equivalent to the Stratonovich SDE

$$dX_t = \alpha f(X_t) dt + f(X_t) \circ W_t$$

which is reducible to

$$dY_t = \alpha dt + dW_t$$

where $Y_t = h(X_t)$ where h is defined as before. Its general solution is

$$X_t = h^{-1}(\alpha t + W_t + h(X_0))$$

STOCHASTIC PARTIAL DIFFERENTIAL EQUATION

Stochastic partial differential equations (SPDEs) generalize partial differential equations via random force terms and coefficients, in the same way ordinary stochastic differential equations generalize ordinary differential equations.

They have relevance to quantum field theory and statistical mechanics.

Examples

One of the most studied SPDEs is the stochastic heat equation, which may formally be written as

$$\partial_t u = \Delta u + \xi \, ,$$

where Δ is the Laplacian and ξ denotes space-time white noise.

Discussion

One difficulty is their lack of regularity. In one space dimension, solutions to the stochastic heat equation are only almost 1/2-Hölder continuous in space and 1/4-Hölder continuous in time. For dimensions two and higher, solutions are not even function-valued, but can be made sense of as random distributions.

AUTOREGRESSIVE MODEL

In statistics and signal processing, an autoregressive (AR) model is a representation of a type of random process; as such, it is used to describe certain time-varying processes in nature, economics, etc. The autoregressive model specifies that the output variable depends linearly on its own previous values and on a stochastic term (an imperfectly predictable term); thus the model is in the form of a stochastic difference equation.

Together with the moving-average (MA) model, it is a special case and key component of the more general ARMA and ARIMA models of time series, which have a more complicated stochastic structure; it is also a special case of the vector autoregressive model (VAR), which consists of a system of more than one stochastic difference equation.

Contrary to the moving-average model, the autoregressive model is not always stationary as it may contain a unit root.

Definition

The notation $AR(p)$ indicates an autoregressive model of order p. The AR(p) model is defined as

$$X_t = c + \sum_{i=1}^{p} \varphi_i X_{t-i} + \varepsilon_t$$

where $\varphi_1, \ldots, \varphi_p$ are the parameters of the model, c is a constant, and ε_t is white noise. This can be equivalently written using the backshift operator B as

$$X_t = c + \sum_{i=1}^{p} \varphi_i B^i X_t + \varepsilon_t$$

so that, moving the summation term to the left side and using polynomial notation, we have

$$\phi(B)X_t = c + \varepsilon_t \,.$$

An autoregressive model can thus be viewed as the output of an all-pole infinite impulse response filter whose input is white noise.

Some parameter constraints are necessary for the model to remain wide-sense stationary. For example, processes in the AR(1) model with $|\varphi_1| \geq 1$ are not stationary. More generally, for an AR(p) model to be wide-sense stationary, the roots of the polynomial $z^p - \sum_{i=1}^{p} \varphi_i z^{p-i}$ must lie within the unit circle, i.e., each root z_i must satisfy $|z_i| < 1$.

Intertemporal Effect of Shocks

In an AR process, a one-time shock affects values of the evolving variable infinitely far into the future. For example, consider the AR(1) model $X_t = c + \varphi_1 X_{t-1} + \varepsilon_t$. A non-zero value for ε_t at say time t=1 affects X_1 by the amount ε_1. Then by the AR equation for X_2 in terms of X_1, this affects X_2 by the amount $\varphi_1 \varepsilon_1$. Then by the AR equation for X_3 in terms of X_2, this affects X_3 by the amount $\varphi_1^2 \varepsilon_1$. Continuing this process shows that the effect of ε_1 never ends, although if the process is stationary then the effect diminishes toward zero in the limit.

Because each shock affects X values infinitely far into the future from when they occur, any given value X_t is affected by shocks occurring infinitely far into the past. This can also be seen by rewriting the autoregression

$$\phi(B)X_t = c + \varepsilon_t \,.$$

(where the constant term has been suppressed by assuming that the variable has been measured as deviations from its mean) as

$$X_t = \frac{1}{\phi(B)}\varepsilon_t \, .$$

When the polynomial division on the right side is carried out, the polynomial in the backshift operator applied to ε_t has an infinite order—that is, an infinite number of lagged values of ε_t appear on the right side of the equation.

Characteristic Polynomial

The autocorrelation function of an AR(p) process can be expressed as

$$\rho(\tau) = \sum_{k=1}^{p} a_k y_k^{-|\tau|},$$

where y_k are the roots of the polynomial

$$\phi(B) = 1 - \sum_{k=1}^{p} \varphi_k B^k$$

where B is the backshift operator, where $\phi(\cdot)$ is the function defining the autoregression, and where φ_k are the coefficients in the autoregression.

The autocorrelation function of an AR(p) process is a sum of decaying exponentials.

- Each real root contributes a component to the autocorrelation function that decays exponentially.
- Similarly, each pair of complex conjugate roots contributes an exponentially damped oscillation.

Graphs of AR(*p*) Processes

The simplest AR process is AR(0), which has no dependence between the terms. Only the error/innovation/noise term contributes to the output of the process, so in the figure, AR(0) corresponds to white noise.

For an AR(1) process with a positive φ, only the previous term in the process and the noise term contribute to the output. If φ is close to 0, then the process still looks like white noise, but as φ approaches 1, the output gets a larger contribution from the previous term relative to the noise. This results in a “smoothing” or integration of the output, similar to a low pass filter.

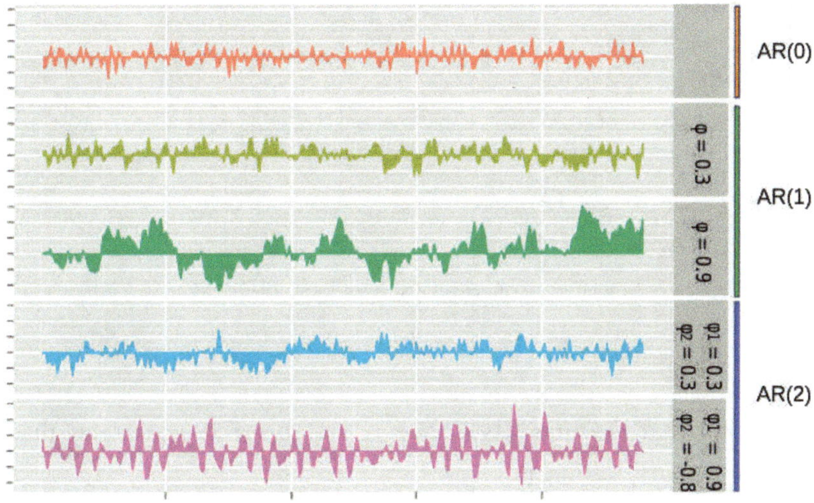

AR(0); AR(1) with AR parameter 0.3; AR(1) with AR parameter 0.9; AR(2) with AR parameters 0.3 and 0.3; and AR(2) with AR parameters 0.9 and −0.8

For an AR(2) process, the previous two terms and the noise term contribute to the output. If both φ_1 and φ_2 are positive, the output will resemble a low pass filter, with the high frequency part of the noise decreased. If φ_1 is positive while φ_2 is negative, then the process favors changes in sign between terms of the process. The output oscillates. This can be likened to edge detection or detection of change in direction.

Example: An AR(1) Process

An AR(1) process is given by:

$$X_t = c + \varphi X_{t-1} + \varepsilon_t$$

where ε_t is a white noise process with zero mean and constant variance σ_ε^2. (Note: The subscript on φ_1 has been dropped.) The process is wide-sense stationary if $|\varphi| < 1$ since it is obtained as the output of a stable filter whose input is white noise. (If $\varphi = 1$ then X_t has infinite variance, and is therefore not wide sense stationary.) Assuming $|\varphi| < 1$, the mean $\mathrm{E}(X_t)$ is identical for all values of t by the very definition of wide sense stationarity. If the mean is denoted by μ, it follows from

$$\mathrm{E}(X_t) = \mathrm{E}(c) + \varphi\mathrm{E}(X_{t-1}) + \mathrm{E}(\varepsilon_t),$$

that

$$\mu = c + \varphi\mu + 0,$$

and hence

$$\mu = \frac{c}{1-\varphi}.$$

In particular, if $c = 0$, then the mean is 0.

The variance is

$$\mathrm{var}(X_t) = \mathrm{E}(X_t^2) - \mu^2 = \frac{\sigma_\varepsilon^2}{1-\varphi^2},$$

where σ_ε is the standard deviation of ε_t. This can be shown by noting that

$$\mathrm{var}(X_t) = \varphi^2 \mathrm{var}(X_{t-1}) + \sigma_\varepsilon^2,$$

and then by noticing that the quantity above is a stable fixed point of this relation.

The autocovariance is given by

$$B_n = \mathrm{E}(X_{t+n}X_t) - \mu^2 = \frac{\sigma_\varepsilon^2}{1-\varphi^2}\,\varphi^{|n|}.$$
{\

It can be seen that the autocovariance function decays with a decay time (also called time constant) of $\tau = -1/\ln(\varphi)$ [to see this, write $B_n = K\varphi^{|n|}$ where K is independent of n. Then note that $\varphi^{|n|} = e^{|n|\ln\varphi}$ and match this to the exponential decay law $e^{-n/\tau}$].

The spectral density function is the Fourier transform of the autocovariance function. In discrete terms this will be the discrete-time Fourier transform:

$$\Phi(\omega) = \frac{1}{\sqrt{2\pi}}\sum_{n=-\infty}^{\infty} B_n e^{-i\omega n} = \frac{1}{\sqrt{2\pi}}\left(\frac{\sigma_\varepsilon^2}{1+\varphi^2-2\varphi\cos(\omega)}\right).$$

This expression is periodic due to the discrete nature of the X_j, which is manifested as the cosine term in the denominator. If we assume that the sampling time $(\Delta t = 1)$ is much smaller than the decay time (τ), then we can use a continuum approximation to B_n:

$$B(t) \approx \frac{\sigma_\varepsilon^2}{1-\varphi^2}\,\varphi^{|t|}$$

which yields a Lorentzian profile for the spectral density:

$$\Phi(\omega) = \frac{1}{\sqrt{2\pi}}\frac{\sigma_\varepsilon^2}{1-\varphi^2}\frac{\gamma}{\pi(\gamma^2+\omega^2)}$$

where $\gamma = 1/\tau$ is the angular frequency associated with the decay time (τ).

An alternative expression for X_t can be derived by first substituting $c + \varphi X_{t-2} + \varepsilon_{t-1}$ for X_{t-1} in the defining equation. Continuing this process N times yields

$$X_t = c\sum_{k=0}^{N-1}\varphi^k + \varphi^N X_{t-N} + \sum_{k=0}^{N-1}\varphi^k \varepsilon_{t-k}.$$

For N approaching infinity, φ^N will approach zero and:

$$X_t = \frac{c}{1-\varphi} + \sum_{k=0}^{\infty}\varphi^k \varepsilon_{t-k}.$$

It is seen that X_t is white noise convolved with the φ^k kernel plus the constant mean. If the white noise ε_t is a Gaussian process then X_t is also a Gaussian process. In other cases, the central limit theorem indicates that X_t will be approximately normally distributed when φ is close to one.

Explicit Mean/Difference form of AR(1) Process

The AR(1) model is the discrete time analogy of the continuous Ornstein-Uhlenbeck process. It is therefore sometimes useful to understand the properties of the AR(1) model cast in an equivalent form. In this form, the AR(1) model is given by:

$X_{t+1} = X_t + \theta(\mu - X_t) + \epsilon_{t+1}$, where $|\theta| < 1$ and μ is the model mean.

By putting this in the form $X_{t+1} = c + \phi X_t$, and then expanding the series for X_{t+n}, one can show that:

$\mathrm{E}(X_{t+n}|X_t) = \mu\left[1-(1-\theta)^n\right] + X_t(1-\theta)^n$, and

$$\mathrm{Var}(X_{t+n}|X_t) = \sigma^2\frac{\left[1-(1-\theta)^{2n}\right]}{1-(1-\theta)^2}.$$

Choosing the Maximum Lag

The partial autocorrelation of an AR(p) process is zero at lag p + 1 and greater, so the appropriate maximum lag is the one beyond which the partial autocorrelations are all zero.

Calculation of the AR Parameters

There are many ways to estimate the coefficients, such as the ordinary least squares procedure or method of moments (through Yule–Walker equations).

The AR(p) model is given by the equation

$$X_t = \sum_{i=1}^{p}\varphi_i X_{t-i} + \varepsilon_t.$$

It is based on parameters φ_i where i = 1, ..., p. There is a direct correspondence between these parameters and the covariance function of the process, and this correspondence can be inverted to determine the parameters from the autocorrelation function (which is itself obtained from the covariances). This is done using the Yule–Walker equations.

Yule–Walker Equations

The Yule–Walker equations, named for Udny Yule and Gilbert Walker, are the following set of equations.

$$\gamma_m = \sum_{k=1}^{p} \varphi_k \gamma_{m-k} + \sigma_\varepsilon^2 \delta_{m,0},$$

where m = 0, ..., p, yielding p+1 equations. Here γ_m is the autocovariance function of X_t, σ_ε is the standard deviation of the input noise process, and $\delta_{m,0}$ is the Kronecker delta function.

Because the last part of an individual equation is non-zero only if m = 0, the set of equations can be solved by representing the equations for m > 0 in matrix form, thus getting the equation

$$\begin{bmatrix} \gamma_1 \\ \gamma_2 \\ \gamma_3 \\ \vdots \\ \gamma_p \end{bmatrix} = \begin{bmatrix} \gamma_0 & \gamma_{-1} & \gamma_{-2} & \cdots \\ \gamma_1 & \gamma_0 & \gamma_{-1} & \cdots \\ \gamma_2 & \gamma_1 & \gamma_0 & \cdots \\ \vdots & \vdots & \vdots & \ddots \\ \gamma_{p-1} & \gamma_{p-2} & \gamma_{p-3} & \cdots \end{bmatrix} \begin{bmatrix} \varphi_1 \\ \varphi_2 \\ \varphi_3 \\ \vdots \\ \varphi_p \end{bmatrix}$$

which can be solved for all $\{\varphi_m; m = 1, 2, \cdots, p\}$ The remaining equation for m = 0 is

$$\gamma_0 = \sum_{k=1}^{p} \varphi_k \gamma_{-k} + \sigma_\varepsilon^2,$$

which, once $\{\varphi_m; m = 1, 2, \cdots, p\}$ are known, can be solved for σ_ε^2

An alternative formulation is in terms of the autocorrelation function. The AR parameters are determined by the first p+1 elements $\rho(\tau)$ of the autocorrelation function. The full autocorrelation function can then be derived by recursively calculating

$$\rho(\tau) = \sum_{k=1}^{p} \varphi_k \rho(k - \tau)$$

Examples for some Low-order AR(p) processes

- p=1
  - $\gamma_1 = \varphi_1 \gamma_0$
  - Hence $\rho_1 = \gamma_1 / \gamma_0 = \varphi_1$
- p=2
  - The Yule–Walker equations for an AR(2) process are

$$\gamma_1 = \varphi_1 \gamma_0 + \varphi_2 \gamma_{-1}$$
$$\gamma_2 = \varphi_1 \gamma_1 + \varphi_2 \gamma_0$$

- Remember that $\gamma_{-k} = \gamma_k$
- Using the first equation yields $\rho_1 = \gamma_1/\gamma_0 = \frac{\varphi_1}{1-\varphi_2}$
- Using the recursion formula yields $\rho_2 = \gamma_2/\gamma_0 = \frac{\varphi_1^2 - \varphi_2^2 + \varphi_2}{1-\varphi_2}$

Estimation of AR Parameters

The above equations (the Yule–Walker equations) provide several routes to estimating the parameters of an AR(p) model, by replacing the theoretical covariances with estimated values. Some of these variants can be described as follows:

- Estimation of autocovariances or autocorrelations. Here each of these terms is estimated separately, using conventional estimates. There are different ways of doing this and the choice between these affects the properties of the estimation scheme. For example, negative estimates of the variance can be produced by some choices.
- Formulation as a least squares regression problem in which an ordinary least squares prediction problem is constructed, basing prediction of values of X_t on the p previous values of the same series. This can be thought of as a forward-prediction scheme. The normal equations for this problem can be seen to correspond to an approximation of the matrix form of the Yule–Walker equations in which each appearance of an autocovariance of the same lag is replaced by a slightly different estimate.
- Formulation as an extended form of ordinary least squares prediction problem. Here two sets of prediction equations are combined into a single estimation scheme and a single set of normal equations. One set is the set of forward-prediction equations and the other is a corresponding set of backward prediction equations, relating to the backward representation of the AR model:

$$X_t = c + \sum_{i=1}^{p} \varphi_i X_{t-i} + \varepsilon_t^* .$$

Here predicted of values of X_t would be based on the p future values of the same series. This way of estimating the AR parameters is due to Burg, and is called the Burg method: Burg and later authors called these particular estimates "maximum entropy estimates", but the reasoning behind this applies to the use of any set of estimated AR parameters. Compared to the estimation scheme using only the forward prediction equations, different estimates of the autocovariances are produced, and the estimates have different stability properties. Burg estimates are particularly associated with maximum entropy spectral estimation.

Other possible approaches to estimation include maximum likelihood estimation. Two distinct variants of maximum likelihood are available: in one (broadly equivalent to the forward prediction least squares scheme) the likelihood function considered is that corresponding to the conditional distribution of later values in the series given the initial p values in the series; in the second, the likelihood function considered is that corresponding to the unconditional joint distribution of all the values in the observed series. Substantial differences in the results of these approaches can occur if the observed series is short, or if the process is close to non-stationarity.

Spectrum

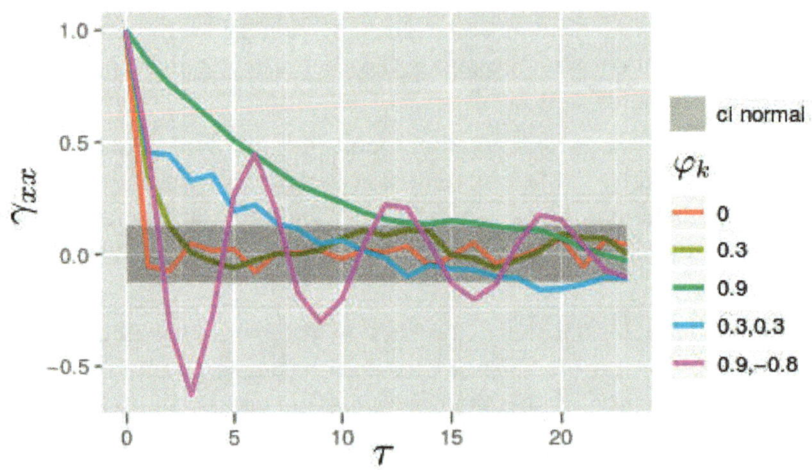

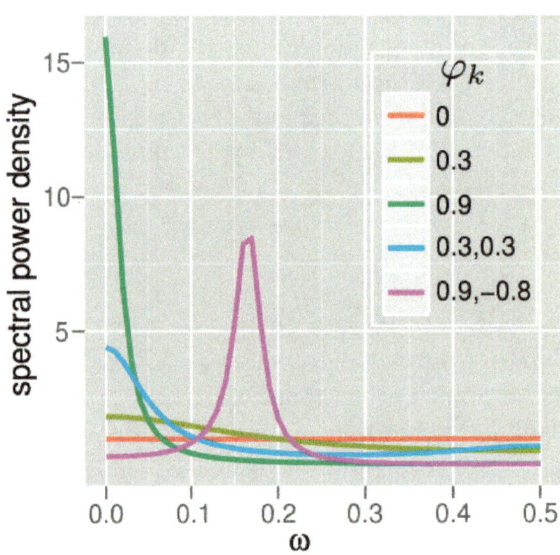

The power spectral density of an AR(p) process with noise variance $\mathrm{Var}(Z_t) = \sigma_Z^2$ is

$$S(f) = \frac{\sigma_Z^2}{|1 - \sum_{k=1}^{p} \varphi_k e^{-i2\pi f k}|^2}.$$

AR(0)

For white noise (AR(0))

$S(f) = \sigma_Z^2.$

AR(1)

For AR(1)

$$S(f) = \frac{\sigma_Z^2}{|1 - \varphi_1 e^{-2\pi i f}|^2} = \frac{\sigma_Z^2}{1 + \varphi_1^2 - 2\varphi_1 \cos 2\pi f}$$

- If $\varphi_1 > 0$ there is a single spectral peak at f=0, often referred to as red noise. As φ_1 becomes nearer 1, there is stronger power at low frequencies, i.e. larger time lags. This is then a low-pass filter, when applied to full spectrum light, everything except for the red light will be filtered.
- If $\varphi_1 < 0$ there is a minimum at f=0, often referred to as blue noise. This similarly acts as a high-pass filter, everything except for blue light will be filtered.

AR(2)

AR(2) processes can be split into three groups depending on the characteristics of their roots:

$$z_1, z_2 = \frac{1}{2}\left(\varphi_1 \pm \sqrt{\varphi_1^2 + 4\varphi_2}\right)$$

- When $\varphi_1^2 + 4\varphi_2 < 0$, the process has a pair of complex-conjugate roots, creating a mid-frequency peak at:

$$f^* = \frac{1}{2\pi} \cos^{-1}\left(\frac{\varphi_1(\varphi_2 - 1)}{4\varphi_2}\right)$$

Otherwise the process has real roots, and:

- When $\varphi_1 > 0$ it acts as a low-pass filter on the white noise with a spectral peak at $f = 0$
- When $\varphi_1 < 0$ it acts as a high-pass filter on the white noise with a spectral peak at $f = 1/2$.

The process is non-stationary when the roots are outside the unit circle. The process is stable when the roots are within the unit circle, or equivalently when the coefficients are in the triangle $-1 \leq \varphi_2 \leq 1 - |\varphi_1|$

The full PSD function can be expressed in real form as:

$$S(f) = \frac{\sigma_Z^2}{1 + \varphi_1^2 + \varphi_2^2 - 2\varphi_1(1 - \varphi_2)\cos(2\pi f) - 2\varphi_2 \cos(4\pi f)}$$

Implementations in Statistics Packages

- R, the stats package includes an ar function.
- MATLAB's Econometrics Toolbox and System Identification Toolbox includes autoregressive models
- Matlab and Octave: the TSA toolbox contains several estimation functions for uni-variate, multivariate and adaptive autoregressive models.

Impulse Response

The impulse response of a system is the change in an evolving variable in response to a change in the value of a shock term k periods earlier, as a function of k. Since the AR model is a special case of the vector autoregressive model, the computation of the impulse response in Vector autoregression#Impulse response applies here.

n-Step-Ahead Forecasting

Once the parameters of the autoregression

$$X_t = c + \sum_{i=1}^{p} \varphi_i X_{t-i} + \varepsilon_t$$

have been estimated, the autoregression can be used to forecast an arbitrary number of periods into the future. First use t to refer to the first period for which data is not yet available; substitute the known prior values X_{t-i} for i=1, ..., p into the autoregressive equation while setting the error term ε_t equal to zero (because we forecast X_t to equal its expected value, and the expected value of the unobserved error term is zero). The output of the autoregressive equation is the forecast for the first unobserved period. Next, use t to refer to the next period for which data is not yet available; again the autoregressive equation is used to make the forecast, with one difference: the value of X one period prior to the one now being forecast is not known, so its expected value—the predicted value arising from the previous forecasting step—is used instead. Then for future periods the same procedure is used, each time using one more forecast value on the right side of the predictive equation until, after p predictions, all p right-side values are predicted values from prior steps.

There are four sources of uncertainty regarding predictions obtained in this manner: (1) uncertainty as to whether the autoregressive model is the correct model; (2) uncertainty about the accuracy of the forecasted values that are used as lagged values in the right side of the autoregressive equation; (3) uncertainty about the true values of the autoregressive coefficients; and (4) uncertainty about the value of the error term ε_t for the period being predicted. Each of the last three can be quantified and combined to give a confidence interval for the n-step-ahead predictions; the confidence interval will become wider as n increases because of the use of an increasing number of estimated values for the right-side variables.

Evaluating the Quality of Forecasts

The predictive performance of the autoregressive model can be assessed as soon as estimation has been done if cross-validation is used. In this approach, some of the initially available data was used for parameter estimation purposes, and some (from available observations later in the data set) was held back for out-of-sample testing. Alternatively, after some time has passed after the parameter estimation was conducted, more data will have become available and predictive performance can be evaluated then using the new data.

In either case, there are two aspects of predictive performance that can be evaluated: one-step-ahead and n-step-ahead performance. For one-step-ahead performance, the estimated parameters are used in the autoregressive equation along with observed values of X for all periods prior to the one being predicted, and the output of the equation is the one-step-ahead forecast; this procedure is used to obtain forecasts for each of the out-of-sample observations. To evaluate the quality of n-step-ahead forecasts, the forecasting procedure in the previous section is employed to obtain the predictions.

Given a set of predicted values and a corresponding set of actual values for X for various time periods, a common evaluation technique is to use the mean squared prediction error; other measures are also available.

The question of how to interpret the measured forecasting accuracy arises—for example, what is a "high" (bad) or a "low" (good) value for the mean squared prediction error? There are two possible points of comparison. First, the forecasting accuracy of an alternative model, estimated under different modeling assumptions or different estimation techniques, can be used for comparison purposes. Second, the out-of-sample accuracy measure can be compared to the same measure computed for the in-sample data points (that were used for parameter estimation) for which enough prior data values are available (that is, dropping the first p data points, for which p prior data points are not available). Since the model was estimated specifically to fit the in-sample points as well as possible, it will usually be the case that the out-of-sample predictive performance will be poorer than the in-sample predictive performance. But if the predictive quality deteriorates out-of-sample by "not very much" (which is not precisely definable), then the forecaster may be satisfied with the performance.

REFERENCES

1. Adomian, George (1983). Stochastic systems. Mathematics in Science and Engineering (169). Orlando, FL: Academic Press Inc.
2. Adomian, George (1986). Nonlinear stochastic operator equations. Orlando, FL: Academic Press Inc.
3. Adomian, George (1989). Nonlinear stochastic systems theory and applications to physics. Mathematics and its Applications (46). Dordrecht: Kluwer Academic Publishers Group.
4. Øksendal, Bernt K. (2003). Stochastic Differential Equations: An Introduction with Applications. Berlin: Springer. ISBN 3-540-04758-1.

5. Teugels, J. and Sund B. (eds.) (2004). Encyclopedia of Actuarial Science. Chichester: Wiley. pp. 523–527.
6. C. W. Gardiner (2004). Handbook of Stochastic Methods: for Physics, Chemistry and the Natural Sciences. Springer. p. 415.
7. Thomas Mikosch (1998). Elementary Stochastic Calculus: with Finance in View. Singapore: World Scientific Publishing. p. 212. ISBN 981-02-3543-7.
8. Seifedine Kadry, (2007). A Solution of Linear Stochastic Differential Equation. USA: WSEAS TRANSACTIONS on MATHEMATICS, April 2007. p. 618. ISSN 1109-2769.
9. Bachelier, L., (1900). Théorie de la speculation (in French), PhD Thesis. NUMDAM: http://archive.numdam.org/ARCHIVE/ASENS/ASENS_1900_3_17_/ASENS_1900_3_17__21_0/ASENS_1900_3_17__21_0.pdf. In English in 1971 book ‹The Random Character of the Stock Market' Eds. P.H. Cootner. External link in |publisher= (help)
10. P.E. Kloeden & E. Platen, (1995). Numerical Solution of Stochastic Differential Equations,. Springer,.
11. Higham., Desmond J. (January 2001). "An Algorithmic Introduction to Numerical Simulation of Stochastic Differential Equations". SIAM Review. 43 (3): 525–546. Bibcode:2001SIAMR..43..525H. doi:10.1137/S0036144500378302.
12. Yule, G. Udny (1927) "On a Method of Investigating Periodicities in Disturbed Series, with Special Reference to Wolfer's Sunspot Numbers", Philosophical Transactions of the Royal Society of London, Ser. A, Vol. 226, 267–298.]
13. Walker, Gilbert (1931) "On Periodicity in Series of Related Terms", Proceedings of the Royal Society of London, Ser. A, Vol. 131, 518–532.
14. Theodoridis, Sergios. "Chapter 1. Probability and Stochastic Processes". Machine Learning: A Bayesian and Optimization Perspective. Academic Press, 2015,. pp. 9–51. ISBN 978-0-12-801522-3.
15. Von Storch, H.; F. W Zwiers (2001). Statistical analysis in climate research. Cambridge Univ Pr. ISBN 0-521-01230-9.[page needed]
16. Burg, J. P. (1968). "A new analysis technique for time series data". In Modern Spectrum Analysis (Edited by D. G. Childers), NATO Advanced Study Institute of Signal Processing with emphasis on Underwater Acoustics. IEEE Press, New York.
17. Brockwell, Peter J.; Dahlhaus, Rainer; Trindade, A. Alexandre (2005). "Modified Burg Algorithms for Multivariate Subset Autoregression" (PDF). Statistica Sinica. 15: 197–213. Archived from the original (PDF) on 2012-10-21.
18. Burg, J.P. (1967) "Maximum Entropy Spectral Analysis", Proceedings of the 37th Meeting of the Society of Exploration Geophysicists, Oklahoma City, Oklahoma.
19. Bos, R.; De Waele, S.; Broersen, P. M. T. (2002). "Autoregressive spectral estimation by application of the burg algorithm to irregularly sampled data". IEEE Transactions on Instrumentation and Measurement. 51 (6): 1289. doi:10.1109/TIM.2002.808031.

20. Mills, Terence C. (1990). Time Series Techniques for Economists. Cambridge University Press.

21. Percival, Donald B.; Walden, Andrew T. (1993). Spectral Analysis for Physical Applications. Cambridge University Press.

22. Pandit, Sudhakar M.; Wu, Shien-Ming (1983). Time Series and System Analysis with Applications. John Wiley & Sons.

Index